LIGHTS A

LIGHTS AND PIGMENTS

Colour Principles for Artists

ROY OSBORNE

John Murray

First published 1980
by John Murray (Publishers) Ltd
50 Albemarle Street, London W1X 4BD

Filmset by Northumberland Press Ltd,
Gateshead, Tyne and Wear
Printed in Great Britain by
Fletcher & Son Ltd, Norwich

British Library Cataloguing in Publication Data
Osborne, Roy
Lights and pigments.
1. Color in art
I. Title
535.6′02′47 N7430.5

ISBN 0–7195–3739–8 (cased)
0–7195–3747–9 (paperback)

Contents

INTRODUCTION *ix*

PART I

1 COLOUR PERCEPTION
The complex nature of colour perception: constancy and association *3*

2 LIGHTS AND PIGMENTS
The spectrum of colour; the additive mixture of lights; the subtractive mixture of pigments *8*

3 SELECTIVE ABSORPTION
The absorption, reflection and transmission of light; colour filters *14*

4 COLOUR SENSATION
Wavelength and colour sensation; rod and cone vision *19*

5 COLOUR APPEAL
The emotional appeal of colour; colour symbolism and phototherapy *29*

PART II

6 LIGHT SOURCES
Incandescent and luminescent sources *37*

7 ADDITIVE COLOUR REPRODUCTION
Television and video; holography *48*

8 PIGMENT SOURCES
Pigments and dyestuffs, natural and artificial *58*

9 VEHICLES AND BINDERS
Pigment binders, natural and artificial 63

10 SUBTRACTIVE COLOUR REPRODUCTION
Colour process printing; photography and film 73

PART III

11 THE MEASUREMENT OF LIGHTS
Photometry, colorimetry and the CIE system 85

12 THE MEASUREMENT OF PIGMENTS
Lightness, hue and saturation; the Ostwald and Munsell systems 95

13 COLOUR INTERACTION
Colour irradiation: contrast and isolation 107

14 OPTICAL COLOUR MIXING
Persistence of vision; optical mixing discs and mosaics 114

15 COLOUR AND FORM
Figure and ground; colour and form 121

PART IV *Appendices*

1 GLOSSARY 133

2 BIOGRAPHIES 139

3 LIST OF PIGMENTS 149

4 SELECT BIBLIOGRAPHY 153

INDEX 157

Illustrations

COLOUR

1.1	The principle of additive colour mixture.	*f.p.* 22
1.2	The principle of subtractive colour mixture.	22
1.3	Selective absorption by coloured surfaces.	22
2.1	Television phosphor triads.	39
2.2	The principle of three-colour printing.	39
3.1	Runge's colour wheel.	70
3.2	Ostwald's colour wheel.	70
3.3	Munsell's colour wheel.	70
4.1	Simultaneous colour contrast.	87
4.2	Successive colour contrast.	87
5.1	Optical mixing discs.	102
5.2	Optical mixing mosaic.	102
6.1	Leonardo da Vinci, *Mona Lisa*. Paris: Louvre Museum.	119
6.2	Pierre Bonnard, *The Bath*. London: Tate Gallery.	119

FIGURES

2.1	Dispersion of white light by a single glass prism.	*p.* 9
2.2	The reconstitution of white light.	9
2.3		11
3.1	The absorption, reflection and transmission of light.	14
3.2	A magnified paint layer.	18
4.1	Diagram of a wave.	20
4.2	Cross-section of the eyeball.	22
4.3	Spectral sensitivity curves for normal vision.	27
7.1	The RCA shadow mask tube (detail).	51
7.2	Recording a hologram.	55
7.3	The principle of hologram reconstruction.	56
10.1	Enlarged half-tone newspaper picture.	77
11.1	A solid angle or cone.	87
11.2	Maxwell's colour triangle and the spectrum locus.	91
11.3	The CIE chromaticity diagram.	93
12.1	Diagrammatic colour wheel.	101

12.2 Runge's colour sphere. *102*
12.3 Ostwald's colour solid; cross-section. *104*
12.4 Munsell's colour tree; cross-section. *105*
13.1 Simultaneous lightness contrast. *108*
13.2 Successive lightness contrast. *110*
15.1 A chessboard design. *125*
15.2 *Homage to the Square* (Albers). *126*
15.3 *Grafias* (Feeley). *128*

Introduction

LIGHTS AND PIGMENTS is a textbook directed primarily at the student of visual art. It seeks to provide a readily accessible introduction to the subject of colour as a whole, intended as an indispensable basis for subsequent artistic activities in painting, printing, dyeing, sculpture, ceramics, fabrics, photography, film, holography, television, video, light art and architecture.

The recent adoption of many new art media calls for an informed teaching approach, geared to the 1980s. This book integrates concepts and methods associated with *lights and pigments* and relates new to traditional art media.

The text is purposely concise to encourage personal initiative. Where necessary the interested reader should obtain additional information by combining experimentation and observation with reference to specialist publications and the work of practising artists and designers.

LIGHTS AND PIGMENTS is in four parts:

PART I outlines aspects of the psychology of colour perception, the physics of colour phenomena, the physiology of colour sensation and the emotional appeal of colour.

PART II catalogues all available colour techniques with a view to helping the artist select media which best suit a particular purpose. The most important sources of lights and pigments are included, together with related methods of colour image reproduction. Following a brief description of each medium, the reader is referred to the work of an artist highly accomplished in its application.

PART III deals with colour appearance: its 'objective' and 'subjective' measurement, the principles of optical contrast and mixing, and the interaction of colour and form. The work of selected artists is discussed in some depth.

PART IV consists of a short glossary of terms, brief biographies of all persons mentioned in the text, a list of pigments in current use and a recommended reading list.

IN ACKNOWLEDGEMENT I would like to thank Ray and Ruby Ovington and Debbie Thorpe, without whom this book would never have been started, and Ruth Kimber and Jim Fryer, without whom it would never have been finished.

I would also like to thank Reg Joyce, Kent Jones, Arnold Myers, Lisa Ridley, Jo Simmonds and all those artists who kindly gave details of their recent work.

PART I

CHAPTER 1

Colour Perception

'We see what is behind our eyes.' Chinese proverb

At the Royal Institution in 1861, the Scots physicist James Clerk Maxwell performed an experiment which demonstrated one aspect of the complex nature of human colour perception.

Using a tartan ribbon as his subject, Maxwell had taken three photographs through three colour filters corresponding to the three primary colours of light. From each of these he made a black-and-white positive transparency (a photographic print on clear glass instead of paper). Harnessing three white-light projectors, he now projected each transparency through its respective colour filter and aligned them to make a single image on a white screen. The resulting image, to the surprise of his audience, reproduced the original design not in black and white but in full colour.[1]

Almost a century after Maxwell's demonstration, the American inventor Edwin H. Land proved that the photographic reproduction of a full-colour scene can be achieved adequately with only two colour sources instead of three.[2] In one example, a pair of black-and-white positive transparencies are made by photographing a scene in turn through red and green filters. Using white-light projectors, the 'red' transparency is projected through the red filter and the 'green' transparency projected without a filter to constitute a single image on a white screen. The resulting image displays a faithful colour reproduction of the original scene (though the perception of some colours may be neither perfect nor immediate).

[1] James Clerk Maxwell (1861), *Experiments on colour as perceived by the eye. Phil. trans.*

[2] Edwin H. Land, *Experiments in Color Vision. Scientific American*, May 1959. Reprint 223.

In 1911, a system of colour cinematography was launched commercially in which red and green filters were passed in front of a camera aperture, in synchronisation with the movement of the film, so that film frames are exposed alternately to filtered red and green light. When the black-and-white film print was projected through a similar filter system, the moving picture appeared multicoloured. Land was aware of early two-colour photographic processes and has observed that when lights are mixed as part of a pattern, and especially when such a pattern represents objects, a far greater wealth of colour is seen than when the same lights are viewed as isolated mixtures. Though one may *sense* patterns of colour one does not as a rule *perceive* patterns of colour. One recognises objects.

In everyday experience, the colour an object appears is not necessarily dependent on the colour of the light which illuminates it. For example, when illumination is weak or coloured the mind may ignore the colours which actually present themselves for viewing and notice nothing remarkable in the appearance of the objects seen. This response to visual stimuli is due to an aspect of *colour constancy* in which colours are perceived not as they are but rather as one may anticipate or expect them to be. The effect appears to derive from an awareness on the part of the observer that the object is a separate entity from the light which falls on to it.

Colour constancy may be a mechanistic response, stemming from inborn perceptual functions, but experience and memory seem also to play their part. The fixed idea or memory of the colours of objects seen in white daylight is very strong for most people and it may be that those who firmly memorise normal colour appearances may subsequently disregard the 'unnatural' colouring of the same or similar objects when their illuminating light is *not* white. Thomas Young had observed, in 1801, that 'when a room is illuminated by the yellow light of a candle, or by the red light of a fire, a sheet of writing paper still appears to retain its whiteness'.

While the capacity to remember the exact quality of colours is generally rather poor, even mildly abnormal changes are immediately apparent when very familiar objects are involved. One notices at once, for instance, if the complexion of a friend differs only slightly from normal—when he or she looks a little 'off colour'. Visual memory can be equally keen when assessing the ripeness of fruit and vegetables. Yet, no matter how familiar one is with their appearance when fresh, the appetite will diminish or even vanish if fresh fruit and vegetables are served in *blue* or *violet* light! Here the mind seems unwilling to ignore or compensate for the unnatural appearance of the objects seen.

In other circumstances, marked alterations in the colour of the illuminating light do not produce unacceptable changes in the colours of objects. This aspect of colour constancy is illustrated in the example of a green lawn partly in direct sunlight and partly in shadow. Despite considerable colour contrast between the two areas—as a painter matching pigments would readily discover—the observer will assume that the lawn is flat, whole, and that both areas are locally the same 'green' colour; the lawn is consequently perceived as such.

Modern techniques of image reproduction, including television, photography and process printing, offer at best only approximate colour rendering. Colour constancy, whether mechanistic, psychological, or a combination of both, is a fundamental part of the highly complex role played by the mind in its response to the incomplete visual information such techniques offer. When faced with visual ambiguity, as in the example of the green lawn, the mind will tend to perceive the most probable interpretation of the information presented.

The strong association between an object and its standard appearance in daylight is one which the artist often strives to weaken. In the mid-nineteenth century, when European painters were exploring a new approach to colour, aspects of colour constancy presented inhibiting factors which often set the artist against his critic. When the English Romantic painter J. M. W. Turner rendered truthful representations of a luminous *yellow* sky at sunset, or the pastel mists of dawn, his contemporaries thought him quite mad; and when Chevreul, the celebrated French chemist, explained how orange sunlight induced *violet* shadows, he was not believed. Yet yellow skies and violet shadows were there for all to see.

The Impressionist painter Claude Monet wished he had been born blind and then suddenly gained his sight as he thought this condition might have freed him from acquired colour associations. Monet wanted to see colours as might a child, unconditioned by visual experience, and so be naïvely sensitive to colour appearances and colour contrasts. This proposal, taken from a reference by Ruskin (1857) to 'our recovery of what may be called the *innocence of the eye*', was to inspire Léger and Delaunay in their own efforts to relieve colour of its representational role. All called on the perception of colour by artist and observer alike to be heightened *not* in respect of the fixed memory of the 'normal' colouring of objects (as seen in white daylight) but to the awareness of the alert colourist who, in order to be true to his model, might render a sky pink, a face blue or a shadow green.

It is a principal difficulty when teaching art to persuade the student to see colours as they really are, that is, free from standard colour associations. Inasmuch as two visual functions can be separately identified, seeing can be said to be accomplished by a combination of acute *sensing* in the eye and correct *perceiving* in the mind. In order to avoid colour constancy responses, or to induce them where necessary, the student should acknowledge a distinction between the 'objective' nature of sensation and the 'subjective' nature of perception. Failure to do this, for instance in photographic portraiture, might result in unacceptable colour rendering in a print which the photographer did not notice and for which the film was not responsible.

To separate objects from their standard colour associations, Monet told his pupils:

> 'Try to forget what objects you have before you—a tree, a house, a field, or whatever. Merely think, here is a little square of blue, here an oblong of pink, here a streak of

yellow, and paint it just as it looks to you, the exact colour and shape, until it gives you your own naïve impression of the scene before you.'[3]

[3] Claude Monet, in an interview with Lilla Cabot Perry. *American Magazine of Art*, March 1927.

CHAPTER 2

Lights and Pigments

> 'White, is a Concentrating, or an Excess of Lights. Black is a deep Hiding, or Privation of Lights. But both are the Produce of all the Primitive Colours compounded or mixed together; the one by Impalpable Colours, and the other by Material Colours.'
>
> Jacob Christoph Le Blon, *Coloritto*, 1735

In 1666, Isaac Newton, then a young man of twenty-three, was the first to study the spectrum of colour scientifically. For this purpose he placed a triangular wedge or *prism* of glass in the path of a beam of sunlight which was allowed to enter a darkened room through a small hole in the window shutter. The rays emerging from the glass prism, when directed to fall on to a white panel, exhibited the vivid light display of the solar spectrum of colour. Newton had shown a simple method by which white light could be broken up into its component colours, a phenomenon known as the *dispersion* of light.

If two isolated light rays, a red and a violet, are projected at the same oblique angle into a glass block, they are affected differently. In a vacuum, all light rays move at the same speed; but in an optically denser medium, such as glass, the speeds of the coloured rays differ: the violet travels more slowly than the red. Because of this difference in speed, the path of the violet rays is bent or *refracted* to a greater angle than that of the red as they pass into the glass block (Figure 2.1).

The red and violet rays form the boundaries of the spectral colour series that can be sensed in the human eye; between them is the colour sequence, orange, yellow, green and blue. In a rainstorm, the round drops of water falling through the air act as the tiny prisms responsible for the brilliant sunlit spectacle of the rainbow. Light from the sun, behind the observer, is refracted and reflected internally in each drop of water and the sunlight dis-

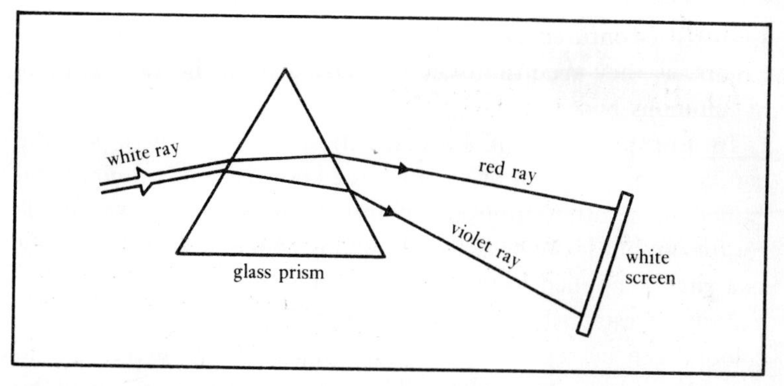

Figure 2.1 Dispersion of white light by a single glass prism.

persed into its full range of component colours, from red on the outer rim of the arc to violet on the inner rim.

When directing coloured rays dispersed by one prism to pass through a converging lens and then through a second prism, Newton found that the emergent beam of light reconstituted the appearance of the original beam of white light (Figure 2.2). From this observation, he deduced that colour is contained neither in the glass nor in the reflecting screen but *in the light itself*. 'Hence therefore it comes to pass,' he wrote (1672) in a letter to the Royal Society, 'that *Whiteness* is the usual Colour of *Light*; for

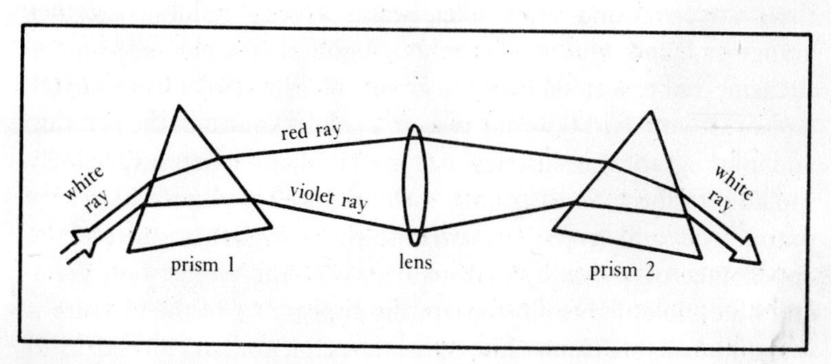

Figure 2.2 The reconstitution of white light.

Light is a confused aggregate of Rays imbued with all sorts of Colors, as they are promiscuously darted from the various parts of luminous bodies.'

In Europe, Newton's investigations completely upset the contemporary concept of colour. As a knowledge of colour is still gained most often from the paintbox, it is easy to share the confusion; but lessons learned in mixing coloured *pigments* cannot as a rule be applied to the mixing of coloured *lights*.

Newton established that white light, a mixture of all spectral colours, can be separated into its components by directing it to pass through a glass prism. The phenomenon is reversed by recomposing or *adding* together the coloured rays, producing mixtures of spectral lights and ultimately obtaining light which appears white to the eye.

If carefully selected, it is found that white light can be obtained using only *three* spectral lights. The three lights, which can be mixed to match a large proportion of all colours appreciable by human colour vision, are *orange-red*, *green* and *blue-violet*.

To observe the superimposition of these lights, three white-light projectors are placed side by side in a darkened room. Each has in front of its lens a filter permitting the passage of one of the three colours. The beams are thrown on to a white, reflecting screen and positioned to overlap, forming the pattern shown in Figure 2.3 (*see* Colour Plate 1.1).

The result of the superimposition is readily observed. Where the orange-red and blue-violet beams overlap (adding together orange-red and blue-violet lights), light of a vivid *magenta* or fuchsine-red is seen. Where the green and blue-violet beams overlap, an area of vivid sky-blue is seen, a colour known in the printing and photography industries as *cyan.* Perhaps most surprisingly (notably to those familiar only with mixtures of pigments), where orange-red and green are mixed, the two lights imitate the appearance of *yellow* light. Finally, where the orange-red, green and blue-violet lights all overlap, the appearance of the mixture is indistinguishable from white light, though it consists physically of three colours only.

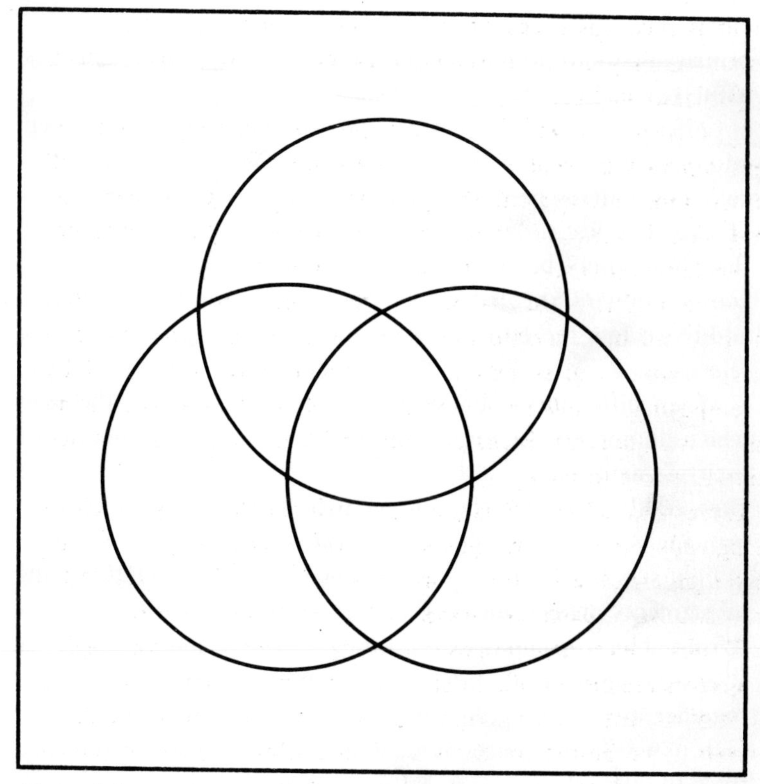

Figure 2.3

Each of the derived colour mixtures—the magenta, cyan and yellow—appears more luminous than either of its colour constituents. This is because, in each case, two sources of light energy are combined or added together. The brightness or intensity of the resulting light is equal to the sum of its components and the mixing of coloured lights in this way is in consequence known as *additive colour mixing*.

The three fundamental colour stimuli, which cannot themselves be imitated by any mixture of coloured lights, are called the additive *primary* colours, namely, orange-red, green and blue-violet. The colours which result from the intermixture of any two primary lights are accordingly called the additive *secondary* colours, namely, magenta, cyan and yellow.

Other than by superimposing lights simultaneously on a screen, additive colour mixing is achieved when a succession of differently coloured lights is presented so rapidly to the eye that perception of flicker is lost; or when adjacent points of light are so small that they cannot be distinguished individually by the eye. It is a combination of the latter two phenomena which permits the additive colour reproduction of an original scene by the medium of television.

When differently coloured *pigments* are intermixed, the light which illuminates the mixture is absorbed in varying amounts by each pigment used. Colours obtained by mixing paints, inks or dyes result *not* from the adding together of light energy but principally from the absorption or *subtraction* by those materials of light supplied by a single or composite source. The intermixing of pigments or dyestuffs is therefore known as *subtractive colour mixing*. This explains why—contrary to the fusion of lights—as a greater number of differently coloured pigments are mixed together, the resulting mixture will darken and appear muddy: as each new colour is introduced, more and more light is absorbed from the illuminating source, which is usually white.

As with mixtures of lights, however, only three primary colours are required in mixtures of pigments in order to match a very large number of colours appreciable by human colour vision. The necessary colours are the familiar paint primaries of red, blue and yellow, or more correctly, *magenta-red*, *cyan-blue* and *chrome-yellow*. The three pigment or subtractive primaries are themselves fundamental and cannot be imitated by any mixture of other pigments.

When subtractive primary pigments are intermixed in the arrangement shown in Figure 2.3 (*see* Colour Plate 1.2), the secondary colours obtained are as follows: magenta and yellow mix to yield orange-red pigment, cyan and yellow mix to yield green, and magenta and cyan mix to yield blue-violet. All three subtractive primaries combine to give pigment which appears black.

Other than by the physical mixture of coloured materials (such as paints on an artist's palette), subtractive colour mixing occurs when colour filters are positioned one behind the other, the colour of the light reaching the eye being determined by the absorption of parts of the illuminating light source by each filter dye. This enables the subtractive colour reproduction of an original scene to be made by colour process printing and colour photography.

CHAPTER 3

Selective Absorption

'If the colour of a surface is indeed the sum of its reflected light, it follows that the colour of that surface might be affected in the quality of the composition of its illumination.'

Michel-Eugène Chevreul,
De la loi du contraste simultané des couleurs, 1839

Material surfaces differ widely in molecular structure, and each surface behaves in a particular way to the light which falls on to it. In principle, of the total amount of light falling on a surface, a certain portion is *absorbed* by the material while the rest is either *reflected* (if the surface is opaque) or *transmitted* (if the surface is transparent). The phenomenon in which this occurs, which is known as the *selective absorption* of light, represents a simplified explanation why material surfaces, including those which are painted or dyed, appear coloured.

White light is obtained conveniently by combining the light of all three additive primary colours, orange-red, green and blue-

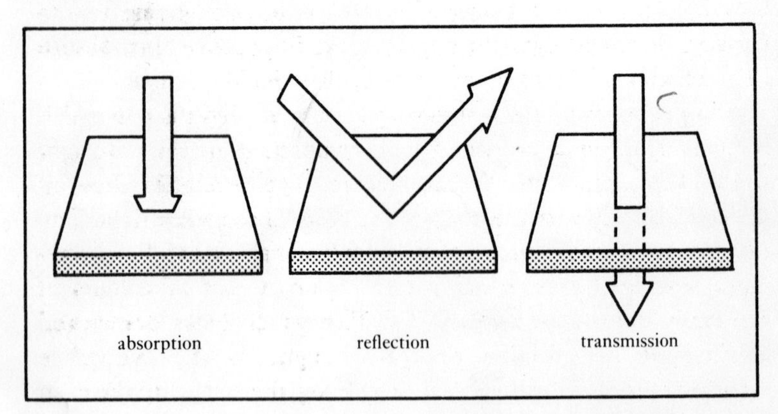

Figure 3.1 The absorption, reflection and transmission of light.

violet. In simple terms, one can visualise that when *one* of these white-light components is wholly absorbed by a surface, the light which remains available for reflection or transmission to the eye of a viewer will consist of the remaining *two*. The surface of a flower petal which looks yellow in daylight is absorbing the blue-violet component of the white light while reflecting both the orange-red and green components, which (by their additive mixture) give the appearance of yellow. Similarly the surface of a leaf which looks green in daylight is absorbing the orange-red and blue-violet components of the daylight while reflecting the green component to the eye (*see* Colour Plate 1.3).

A surface which reflects all three additive primaries, and neither absorbs nor transmits any incident daylight, will appear white and opaque. In practice, an ideal white is nearly obtained with a magnesium oxide smoke coating newly deposited on cold metal, which absorbs about 2 per cent of its incident light; magnesium carbonate absorbs about 11 per cent, while other pigments identifiable as 'white' may absorb up to 20 per cent.

A material which absorbs all three additive primaries, and neither reflects nor transmits any light, appears perfectly black and opaque. In practice, the darkest pigment, Carbon black, absorbs about 97 per cent of the light falling on it. When the three primaries are absorbed uniformly by an opaque surface but only in part, there is a *general absorption* of light giving rise to surfaces which in daylight appear grey; in practice they absorb between about 20 and 90 per cent of their incident light.

An opaque surface which looks yellow in daylight is one which is reflecting the orange-red and green components of the white light source while absorbing the blue-violet. The reflected yellow, if mixed additively with the blue-violet, would make white light. The two colours thereby complete or 'complement' one another, forming a so-called *complementary colour* pair or pair of colours of maximum contrast. Similarly a surface which looks orange-red in daylight is reflecting orange-red light to the eye while absorbing green and blue-violet, which together make up the cyan complementary of the orange-red. A surface which looks green is absorbing orange-red and blue-violet light, which make the magenta complementary of the green.

The physical intermixture of a pair of complementary colours always results in colour cancellation: complementary *lights* combine additively to yield white light; complementary *pigments* combine subtractively, absorbing all incident light, to yield pigment which looks black. When using lights and pigments, it may be helpful to visualise a coloured surface as one which, while reflecting or transmitting the light of its apparent colour to the eye, absorbs the light of its complementary colour from a white light source.

When the light of an illuminating source is *not* white, an important principle of selective absorption remains true: a surface can only reflect or transmit a colour which forms a component of the light source. The portion of the light absorbed by the surface is transformed into heat; the rest remains available for reflection or transmission to the eye of the viewer. For example, in a darkroom lit only by a red safelight, a sheet of 'white' photographic paper (if uninfluenced by colour constancy) looks *red*. This is because both green and blue components of the white light source (an electric lamp) are absorbed by the red dye of the filter and their transmission is prevented. The red rays pass freely through the filter to the paper which, in turn, reflects them to the eye.

In the red illumination given by the darkroom safelight a sheet of paper which looks *blue* in daylight will appear black. This is because the blue component of the white light emitted by the lamp behind the filter is absorbed by the red filter dye: blue light is prevented from reaching the sheet of 'blue' paper which therefore reflects no light. For the same reason a red filter placed over a camera aperture has the optical effect of 'darkening' a blue sky in a black-and-white photograph; white clouds are emphasised as they convey some red light, which the red filter transmits.

In practice, the light reflected or transmitted by a coloured paint, ink or dye viewed in daylight is spectrally never very pure. It almost always consists of a mixture of several spectral colours, only *one* of which predominates and gives to the surface its characteristic colour appearance. As Chevreul observed (1839):

> 'It must not be supposed that a red or a yellow body reflects only *red* and *yellow* rays besides white light; they each reflect *all kinds* of coloured rays; only those rays which lead us to judge the bodies to be *red* or *yellow*, being more numerous than the other rays reflected, produce a greater effect.'

If pigments were pure enough spectrally to reflect the light of a single colour only, one would be faced with the fact that the mixing together of two pigments to yield a third or resultant colour *would not be possible*. For example, in white light, blue and yellow paints mix to make green paint. This is well known but its explanation is not obvious.

A blue pigment, while predominantly reflecting the blue component of a white light source, also reflects some *green* and *violet* light plus smaller amounts of other colours present in the white light. This can be proved by looking at a blue-painted surface in turn through a green filter and a violet filter. If the reflected light contained no trace of either colour, the surface would in each case appear perfectly black (since no light would be transmitted by either filter dye). The surface, however, will appear respectively green and violet. Similarly, a yellow pigment, while predominantly reflecting yellow light, also reflects substantial amounts of *red* and *green*.

From this observation it can now be deduced that green is the colour obtained when blue and yellow paints are mixed because it is the only colour reflected to a significant degree by both the blue and the yellow pigments. When mixed on the palette, all light is absorbed from the white light source—some by the blue pigment and the rest by the yellow—except for the green, which is reflected.

A colourless pane of glass transmits all the light which falls on to it and looks wholly transparent. If the same piece of glass is broken up and ground into a fine powder it will appear white.

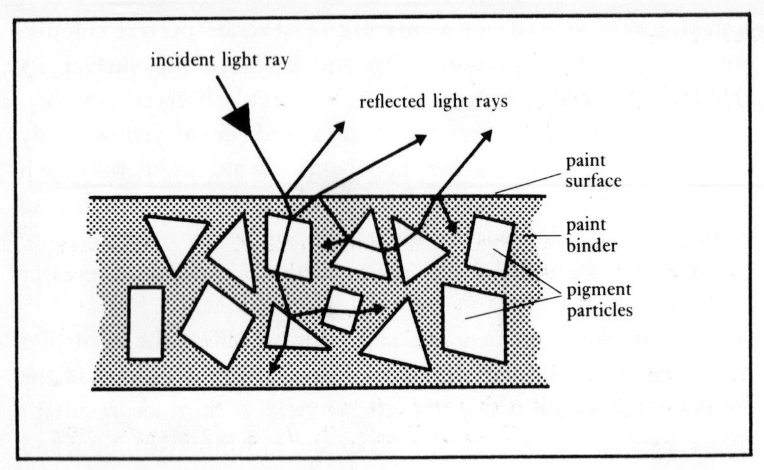

Figure 3.2 A magnified paint layer.

The multiple surfaces of the tiny glass fragments will have the effect of scattering most of the incident light and, if very finely ground, a thin layer of glass powder will be sufficient to obscure any surface on which it is laid. It follows that a 'white' paint or ink layer which consists of such transparent pigment particles will tend to adopt the colour of its illuminating light and appear white in daylight or red under a red safelight.

Most paints and inks consist of finely ground pigment particles held within a transparent, colourless binder. When examined under a microscope, the pigment particles are usually found to be transparent and as such act as individual filters. On entering a layer of paint or ink, light from the illuminating source, intercepted by the pigment, is diffusely scattered and prevented from passing straight through. Each particle absorbs a tiny portion of the incident light while reflecting or transmitting light to neighbouring filter particles. After multiple scattering within the layer, a suitably modified beam of light is reflected to the eye of the viewer and interpreted by him as a single colour impression.

CHAPTER 4

Colour Sensation

'Indeed, Rays properly expressed, are not coloured. There is nothing else in them but a certain Power or Disposition which so conditions them that they produce in us the Sensation of this or that Colour.'

Sir Isaac Newton, *Opticks*, 1704

'We . . . must therefore call light an influence capable of entering the eye and of affecting it with a sense of vision.'

Thomas Young, *On the theory of optics*, 1801

Light is described most conveniently in terms of its apparent properties and the effects it produces. In certain circumstances radiant solar energy, a tiny proportion of which is visible as light, exhibits behaviour typical of *wave* motion.[1]

A wave has four associated qualities: the *velocity* or constant speed at which it moves through a particular medium, the *frequency* of complete waves passing a fixed point per second, the *wave-length* as measured between two adjacent wave crests, and *amplitude*, which is the displacement from zero position (or equilibrium) experienced by a particle of a medium as a wave moves through it.[2]

A familiar example of wave motion is the disturbance which occurs when a stone is thrown into a pool of water. In each complete wave radiating from the point of impact of the stone, a crest occurs where the water surface is in a state of pressure and a trough occurs where the surface is in a state of tension. The energy required to start the water moving is given by the impact of the stone which, although it continues to fall after striking

[1] Radiant solar energy is believed to consist of a continuous electrical and magnetic wave travelling as a discontinuous series of quanta or tiny 'energy packets'.

[2] For numerical values of the velocity, frequency and wavelength of light waves see Appendix 1 (Glossary).

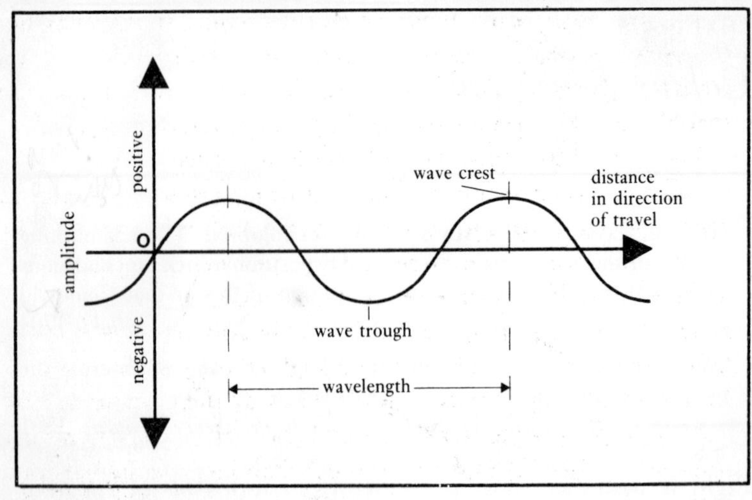

Figure 4.1 Diagram of a wave.

the surface of the water, does so with less speed. Part of the kinetic energy of the falling stone is transferred to the surface of the water where it is responsible for the generation of the wave.

In the human eye the sensation of vision is excited when radiant waves of light energy strike the inner surface of the eye as, in analogy, moving water waves ultimately lap the side of the pool. As a general rule, different frequencies at which successive waves are received give rise to sensations of different colours. Light waves can be specified by frequency or wavelength (the frequency of a complete wave being inversely proportional to its wavelength) though wavelength is the more often used.

Only a small proportion of all solar energy is luminous, that is, capable of arousing the sensation of light in the human eye. The wavelengths of visible light, measured customarily in nanometres,[3] are limited approximately to the tiny waveband 380 to 760 nanometres—about an octave in musical terms. The extent of

[3] The *Systême Internationale* (SI) unit of length equal to one thousand-millionth of a metre.

this waveband of visible light, which corresponds to the spectrum of colour, is not perfectly definable since it depends both on amplitude (the amount of energy carried by a wave) and on the variable nature of colour sensing between different individuals.

Energy of wavelength 760 nanometres is normally sensed in the eye as red light, and shorter wavelengths in turn identify orange, yellow, green, blue and violet light. Beyond either end of the visible colour series, invisible energy 'below' the red, and extending from 760 nanometres to 1 millimetre approximately, is known by the term *infrared* and is readily sensed by the body as heat. 'Above' the violet, invisible energy of the waveband 380 to 10 nanometres approximately is identified by the term *ultraviolet*, and the longer wavelengths of this band (those responsible for sun-tanning) are capable of forming vitamin D in human skin tissue.

Colour sensing is accomplished by a complex optical system in which patterns of light are initially focused by a single-lens mechanism on to the *retina* of the eye. The retina, a fine 'net' of interconnected nerve cells, lines the entire inner surface of the eyeball with the exception of the pupil, the aperture through which light enters. On the retina, light energy is transformed into neural signals, partly electrical and partly chemical in their nature, which are transmitted rapidly along the fibres of the optic nerve and through the mid-brain to the occipital lobes of the cerebral cortex. In this posterior region of the brain the signals, after decoding, form cohesive 'pictures' which not only serve as a basis for immediate action but are also stored in the memory to await future recall.

Greatest sharpness or acuity of vision occurs when light from an illuminated object falls on the *fovea*, an area about 2 millimetres across which defines the visual axis of the eyeball (Figure 4.2). The visual mechanism normally aligns the fovea with whatever is of immediate interest to the observer. The visual field of the fovea is limited to a scanning area about 10 centimetres across at a distance of 2 metres, and the eye shifts constantly to ensure that the focused image falls precisely on the fovea.

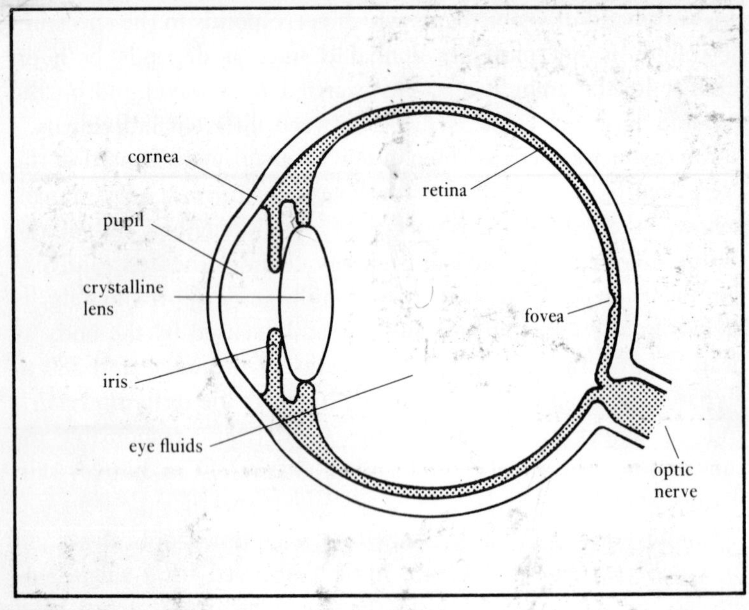

Figure 4.2 Cross-section of the eyeball.

Snatched, fragmented aspects of the overall visual field are duly synthesised by the brain into recognisable forms.

At a suitable reading distance from the eye—33 centimetres or about 14 inches—the eye's field of view comfortably encompasses the page of this book. The area seen sharply, however, is only about 1 centimetre across; and greatest acuity is confined to a single letter at the centre of that area. In order to sense every part of the page when reading, the eye must make an enormous number of tiny shifts, known as *saccades*, which are an important part of visual sensing. The visible world is not normally sensed by a fixed, all-encompassing stare, as Renaissance draughtsmen would have one believe, but rather by the mental synthesis of many momentary glimpses, implied perhaps more closely by Cubist painting and sculpture early in this century.

The retina itself is an 'extension' of the brain which has become sensitive to light, and typical brain cells link the visual

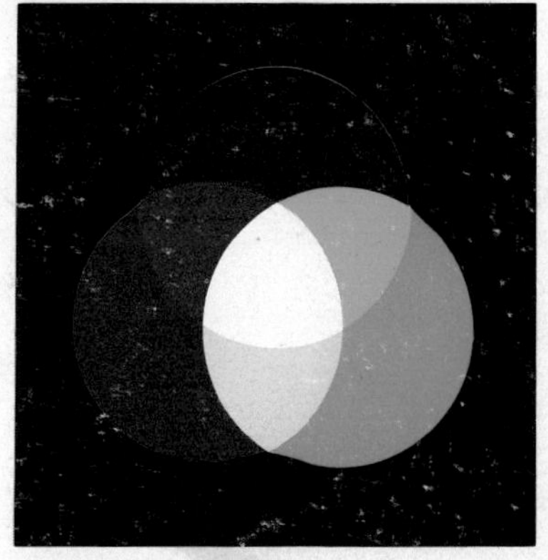

Colour 1.1 The principle of additive colour mixing

Colour 1.2 The principle of subtractive colour mixing

Colour 1.3 Selective absorption by coloured surfaces

receptors of the retina to the fibres of the optic nerve. The two types of receptor which equip the retina are the *rod* and *cone* cells, so named because of their relative shape when observed through a microscope: the rods are straight and thin and the cones generally more bulbous.

Rod and cone receptors are embedded in the lining of the retina remote from the pupil, and to reach them light passing through the cornea, crystalline lens and eye fluids must also penetrate blood vessels and layers of nerve cells involved in the transmission of visual signals to the brain. The rod cells, an estimated 120 million of which are present on each retina, sense light-to-dark values in dim illumination (and are therefore most useful at night) but do not distinguish colours. The cones are sensitive to the different colours of the spectrum and seem to be wholly responsible for the faculty of the eye to sense colour in daylight.

Some seven million cones populate the retina, 100,000 of which are packed into the central fovea, which is rod-free. The fovea not only possesses a concentration of cone cells but also a one-to-one connection between each of its cones and a bipolar nerve cell, the first link in the channel of signal transmission from the retina to the brain. By contrast, on the periphery of the retina, several hundred rods may converge on to a single bipolar cell. This no doubt accounts for the high visual resolution which is characteristic of foveal cone vision but absent from rod vision, which sacrifices acuity in favour of extreme visual sensitivity in semidarkness.

What happens when light energy stimulates the rod or cone receptors is unknown in its entirety. No single theory yet explains cohesively how visual information is encoded on the retina and carried to the brain, nor how, once there, the information is decoded. After leaving the retina, it is obvious that the electrical energy involved can no longer be described in terms of light. The light, having activated specific chemical changes, ceases to exist as such; it has been put to work just as the energy of a released firing pin is transformed in the moment it strikes the bullet.

Investigators agree that when light is sensed by a visual receptor, the initial reaction is photochemical. Rod cells are known to contain *visual purple*, a light-sensitive pigment located in the tip of each rod, which on exposure to light sets off electrical impulses in the adjoining nerve cells. In darkness a large amount of visual purple accumulates in the rods, affording the eye hypersensitivity in weak illumination. On exposure to bright light the visual purple decomposes and the rod cells disfunction.

Visual purple, an organic compound formed with the aid of vitamin A, was extracted from rod cells over a century ago; yet only fairly recently has significant progress been made in identifying the cone-cell pigments. Physiologists speculated that various substances sensitive to different wavebands of light were contained in the cones but no laboratory evidence of their specific nature existed until 1964, when two teams of American scientists succeeded independently in analysing the absorption properties of cone pigments in isolation.[4]

Their observations indicated that cone pigments are of three main types, each differing from the other two in its sensitivity to coloured light: one appears to be responsible primarily for sensing the longer (red) wavelengths of the spectrum, another for sensing the middle (green) wavelengths, and the third for sensing the shorter (blue) wavelengths. The discovery lent support to a theory of three-colour pigment vision which predicted that the retina possesses three types of cone pigment segregated completely or apparently so into three types of cone receptor cell.

A theory of three-component colour vision had first been proposed by Thomas Young. Having established that mixtures of three primary lights yield impressions of a very large number of other coloured lights when projected on to a white screen, Young suggested (1801) that the retina need possess only three varieties of colour-sensitive receptor in order to sense a correspondingly large number of colours in the visible world.

In the 1860s, the German physiologist Hermann von Helmholtz

[4] Edward F. MacNichol, Jr., *Three-Pigment Color Vision, Scientific American*, December 1964. Reprint 197.

revised and clarified Young's theory, confining his experimental approach similarly to the sensing of colours viewed in isolation. It should be remembered however that in everyday life colours are not generally presented as isolated stimuli: they more often identify objects.

In a series of experiments since 1954, Edwin Land has made a methodical study of visual response in the context of coloured pictures and patterns. 'The eye,' he observes, 'has evolved to see the world in unchanging color, regardless of always unpredictable, shifting and uneven illumination.' The hallucinatory changes which colours appear to undergo as a consequence of their close juxtaposition has been studied since the time of Young; Land insists that, if attempting to formulate a comprehensive theory of colour vision, important colour constancy effects must also be taken into account.[5] He has suggested that, rather than responding directly to the physical flow of light energy received, the retinal cone cells encode colour information as a 'three-part report' based on the visually estimated amounts of light reflected from the surface of objects, a response which undermines the over-simplified assumption that the quality of a colour seen corresponds invariably to the wavelength of the light intercepted by the retina.

There are two other, physiological factors which are important to the visual artist in an appreciation of colour sensing by the eye. These are the degree of divergence from 'normal' colour vision, in which 'the eyes of some individuals are not able to distinguish as many colours as those of ordinary persons' (Helmholtz), and the adaptable response of the eye to bright or dim light, in which the sensitivity of the eye to colour is modified as the level of illumination changes.

The consequences of abnormal or defective colour vision may be minor, as the puzzlement if, when a friend adjusts the colour control of a television picture, he adds too much green; yet when someone else adjusts the colour balance, he may insist

[5] Edwin H. Land, *The Retinex Theory of Color Vision, Scientific American*, December 1977.

the colour cast is too red. There are of course much more hazardous implications, as the danger present at night when a motorist with defective colour vision is unable to distinguish the red light from the green of a traffic signal.

The colour sensing of an estimated 8 per cent of men is anomalous; in women, colour vision defects are relatively rare (less than 0·5 per cent). The most common deficiency results in confusion between the sensing of red and green stimuli, which may appear indistinguishable. In extreme colour blindness (monochromatism), the visible world may appear as lighter and darker values of one colour only or of white, black and grey only.

Defective colour vision, which can occur in one eye only, is usually congenital but can be caused by age, disease, drug toxicity, or prolonged exposure to high levels of sound. Long periods of exposure to abnormally bright light also present a danger, the eye being several hundred times more susceptible to damage from light at the violet end of the spectrum than to light at the red end.

For the same expenditure of light energy at each wavelength, human colour vision is not equally sensitive to different colours. It will be noticed that the colours of the spectrum do not exhibit their greatest strength of colour at the same level of apparent brightness: as a rule, the wavelengths in the yellow region of the spectrum look considerably brighter than those in the red, green or blue regions. This lack of correspondence between stimulus and response (first investigated by Bouguer) can be represented in a graph on which wavelength of a spectral stimulus is plotted against the amount of light the stimulus *appears* to convey to the eye. The curve for the whole spectrum is known as a *spectral sensitivity curve*.

As two types of receptor cell are present on the retina there are two sensitivity curves; those shown in Figure 4.3 indicate average response of a large sample of observers. When the cone cells are operative, peak visual sensitivity is found to occur in the yellow region of the spectrum at approximately 555 nanometres. For the same amount of energy expended in physical terms, spectral lights of wavelength 510 (green) and 610 nanometres (orange)

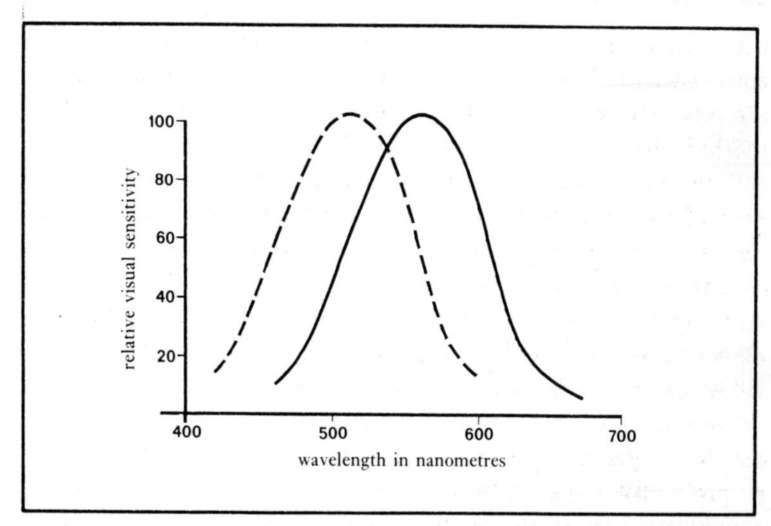

Figure 4.3 Spectral sensitivity curves for normal vision. The continuous line represents the curve for fully light-adapted vision and the dotted line that for fully dark-adapted vision.

appear half as bright; those of 470 (blue) and 650 nanometres (red) appear only one-tenth as bright.

As the level of illumination changes, the capacity of the eye to sense colour is variable. While fully light-adapted (photopic) vision is most sensitive to yellow light, peak visual sensitivity in fully dark-adapted (scotopic) vision shifts to green light in the region of 507 nanometres. This mechanistic adaptation of the eye, which occurs as darkness approaches, is known as the *Purkinje shift*, named after the Czech physiologist who documented the effect early in the nineteenth century. It is of some importance in the practical application of colour since, if a warning signal is to achieve greatest visibility, it should be of the colour which appears most luminous to the eye; this is yellow in daylight and green at night.

Rod and cone receptors adapt at different rates. Rod adaptation in semidarkness may continue for an hour or more whereas cone adaptation to bright light is complete in about seven minutes. The rods are relatively insensitive to red light but highly sensitive to blue light. Thus, as Purkinje had observed, if red and blue surfaces are viewed side by side in dim light, after several minutes, during which time the eye commences its adaptation, the 'red' surface will lose its colour and look black while the blue surface retains its blue appearance. Similarly, at twilight, red objects appear to darken fairly rapidly while blue ones often remain identifiable after the visual discrimination of all other colours is no longer possible.

At night, moonlight may provide rod vision with enough light to see shadows, and therefore perceive objects, but rarely does it give enough light to stimulate the cones and arouse the sensation of colour.

CHAPTER 5

Colour Appeal

> 'Yellow can express happiness, and then again, pain. There is flame red, blood red and rose red. There is silver blue, sky blue and thunder blue. Every colour harbours its own soul, delighting or disgusting or stimulating me.'
>
> Emil Nolde, 1942

Colour is popularly experienced as a conveyor of feeling and, in this century, many European and American artists have reaffirmed a belief in the transmission of feeling through colour. This trend—which developed from Gauguin and Van Gogh's Japanese-influenced break with descriptive colour in painting—led to violently expressive imagery initially by Munch, Nolde, Matisse, Kandinsky and Kirchner. 'Colours, while they can serve to describe things or natural phenomena, have in themselves, independently of the objects they set out to express, an important action on a spectator's feelings,' wrote Matisse (1951); knowing this, the alert colourist can induce in an observer predictable responses of tranquility, excitement or deep anxiety.

One speaks of a *white* lie, a *black* look, of being *green* with envy or feeling *blue*; a world without colour—a world turned *grey*—is depressive and conformist. Though these and other associations are widely recognised, colour also gives rise to important *personal* associations.

'Colour tests' are based on the premise that the personal attraction of one colour and the rejection of another have a specific interpretation in psycho-analytical terms. In such a colour test, as that devised by Max Lüscher,[1] the participant is asked to arrange small, isolated colour samples in order of personal preference and without aesthetic consideration. The results can give a surprisingly

[1] Max Lüscher (1948), *The Lüscher Color Test*. New York: Random House Books, 1969; London: Jonathan Cape, 1970; Pan Books, 1971.

accurate description of personality with indications of possible talents, weaknesses, and emotional and rational states at the time of the test.

Blue and yellow, the two primal colours recognised by Lüscher, form the principal axis of the psychological colour theory of Johann Wolfgang von Goethe, the German poet and dramatist. *Blue*, which is allied to darkness and night, induces inactivity and quiescence; *yellow*, associated with white and the light of day, incites expectation and preparation for activity. According to Goethe and Lüscher, blue and yellow represent the heteronomous or involuntary forces which influence man externally, while the two subordinate colours, red and green, influence the autonomous or voluntary forces. *Red* excites the physical system to attack and conquest; *green* retreats in defence and self-preservation.

Goethe proposed (1820) that 'in order to experience fully these important individual effects, the eye should be entirely surrounded by one colour; we should be in a room of one colour, or look through a coloured glass'. Barnett Newman, in the first exhibition of his large-scale paintings (1949), similarly felt that viewers should stand close to the surface of each work in order that their vision be saturated with colour, as had his own while painting them.

To the occultist, the colour preferences exercised when a new set of clothes, a car or a decorative colour scheme is chosen are believed to relate to the presence of a field of force or influence emanating from each living being. The portion of this influence capable of stimulating vision can be perceived by some observers as a coloured aureole or *aura* which has been interpreted by them as evidence of a noncorporeal 'body' which permeates the physical body and extends outwards from it in all directions.

Between 1939 and his death in 1978, Semyon Kirlian developed optical instruments for photographically documenting various aural effects. He had found that when a living object, such as a hand, is made to act as one of a pair of electrodes, its field of radiating energy will interact with that of a high-frequency electrical current. The resulting glow or 'corona discharge', recorded on photographic paper or film, can then be analysed by studying its relative shape and colour distribution.[2]

The American psychic Edgar Cayce could not remember a time when, visually unaided, he did not see auras emanating from the persons he met, 'with blues and greens and reds gently pouring from their heads and shoulders'.[3] He felt that an easy way to 'figure out' the colours which constitute an aura, for those who do not see them directly, is simply to take note of the colours which individuals wear most often or with which they habitually surround themselves.

Red is traditionally associated with the strong temperaments which mark the most physical human passions: infatuation, bravery, hatred and revenge. Bright red is the colour of the extrovert, active, ambitious and practical; and pink often indicates impulsiveness and immaturity, characteristic of the young. Bright *orange* symbolises vitality and physical endurance, while *yellow*, symbolising the virtues of wisdom and knowledge, is said to stimulate the mind.

Green represents regeneration, growth and the 'salad days' of youthful inexperience. It is nature's colour, reflecting the light rays of the sun in their most soothing form. Cayce notes that green is rarely a dominating colour in the aura: if mixed with blue it indicates honesty and potential powers of healing; if mixed with yellow, its nature is dubious, Ruskin associating dull yellow-green with 'the worst general character that colour can possibly have'.

Blue traditionally symbolises piety, prayer and contemplation, pale blue reflecting the trust and innocence typical of the very young. For Cayce, those who favour deep blue 'have found their work': they have a mission which they steadfastly try to fulfil. *Indigo* and *violet*, the two remaining colours, are often linked with clairvoyance and psychic sensitivity; indigo is associated with spiritual seeking and self-mastery, and violet with intuition and idealism.

[2] Brian and Marita Snellgrove (1979), *The Unseen Self*. Carshalton, Surrey: Brian J. Snellgrove.

[3] Edgar Cayce (1943), *Auras*. Virginia Beach, Virginia: A.R.E. Press, 1973.

Finally, in equilibrium of body, mind and spirit, the spectral colours fuse to yield the *white* halo or nimbus customarily associated with holy figures in pictorial representation.

It has been proposed that the seven spectral colours distinguished in natural philosophy and occultism each possesses distinctive symptomatic or remedial attributes capable of influencing the physical system of the body. A principal belief of the 'colour healer' is that, for each major organ of the human body, light of a specific colour is able to stimulate or inhibit its functioning. Some 2,500 years ago Pythagoras applied coloured light therapeutically and 'colour halls' were used for healing in Ancient Egypt, China and India.

The pioneer of modern colour therapy was Niels Finsen of Denmark. Following the discovery, in 1877, of the bactericidal action of solar ultraviolet energy, Finsen studied the possibility of assisting the healing of wounds with visible light. He subsequently used red light to inhibit the formation of smallpox scars and, in 1896, founded a Light Institute (now the Finsen Institute of Copenhagen) for the phototreatment of tuberculosis.

Fernand Léger, Rudolph Steiner and Charles Edward Iredell later became enthusiastic participants in the search for a 'cure through colour'. Léger proposed 'green and blue wards for the nervous and sick, others yellow and red to stimulate the depressed and anaemic'; Steiner lectured the medical profession praising the virtue and safety of colour healing in harnessing natural defences within the body; and Iredell, consulting surgeon in actino- and radio-therapy at Guy's Hospital, London (1907–37), claimed many cures using coloured lights only, either singly or in sequence.

In the 1970s the laser gained justifiable recognition as an important new medical aid. As a surgical tool it can be used to remove malignant tumours by burning away diseased material while simultaneously sealing surrounding blood vessels. In other areas of medicine it has been found that very low-power laser sources, the heat effects of which are less than 0·001 of a degree Celsius, have proved beneficial in accelerating the healing of various types of skin-tissue damage. Weak laser beams are also used in acupuncture, in place of traditional gold or silver needles, to stimulate those *loci* on the skin's surface at which changes of electrical resistance are detectable.

The physiological influence of colour in one's surroundings, discussed at some length by Goethe (1810), is readily tested personally. In red illumination, and to a lesser extent orange or yellow light, one will tend to experience an increase of muscular tension. Various hormones are released into the bloodstream, including adrenalin, which constricts the blood vessels and steps up blood pressure and heartbeat. Increased oxygen intake (the consequence of a faster rate of respiration) coupled with the necessary increase of haemoglobin in the blood cells in turn raises body temperature.

Conversely, on exposure to bright green or blue light, one will tend to experience a release of muscular tension, a slowing of heartbeat and slight lowering of body temperature. After a period, which may vary between different individuals, the activity promoted by a specific colour is reversed: blood pressure increased by red light becomes more than normally depressed and blood pressure lowered by blue light will subsequently register an abnormal rise.

While the major function of the visual system would appear to be the sensing of light to distinguish objects, a number of retinal nerve endings have been isolated whose function seems *not* to be related primarily to vision. A network of fibres leads directly from the retina to the spinal cord. Information about the nature of the light intercepted by the eye is ultimately channelled back to the pineal and pituitary glands, and it would appear that the presence of nonvisual cells on the retina may represent a photobiological trigger which supplements the chemical stimulation of hormone activity from within the body.

It has been found also that the retina is unable to retain all of the vitamin A which is liberated in the eye during light adaptation. Excess vitamin A is diffused into neighbouring tissues and into the bloodstream. This would appear to connect retinal activity with vitamin A circulation and supports further a possibility that the eye is not only the organ of sight but plays a significant role in the distribution of organic substances throughout the body.

Colour vision can provide information about objects which aids their identification, description and position in space; indeed this may be its principal biological function. Evidence exists which proves that colour also influences thought and feeling: dark objects appear nearer, heavier and smaller than white objects of the same physical dimensions; and candy colours, which look highly attractive in confectionery, would upset the consumer if used to dye savoury foodstuffs. Though scientific discoveries are increasingly supporting aspects of colour experience which both artist and occultist have embraced intuitively for centuries, the experience of colour in its fullest sense may require an openmindedness which may at times run counter to conventional belief.

PART II

CHAPTER 6

Light Sources[1]

Fire has fascinated man ever since he first learned to control it for his own purposes, and firelight has recently been revived as an art medium.

In 1961, Yves Klein installed his *Fire Wall* at the Museum Haus Lange in Krefeld, Germany. This consisted of a large grid of metal pipes supporting a double-sided system of gas jets displaying 100 'International Klein Blue' flames. In the same year, Bernard Aubertin became interested in the art of pyrotechnics and has since fabricated a series of *Spectacles pyromaniaques*, many involving the ignition of constructions of household matches. Two centuries earlier, Louis-Bertrand Castel had employed candlelight in his *Claveçin oculaire* (1734), a five-octave harpsichord keyboard which operated the raising and lowering of translucent coloured tapes illuminated from behind by candle flames.

The emission of light by a material solely because it is heated, which occurs when thermal energy of motion is transformed into light energy, is known as *incandescence*. An incandescent solid, such as a carbon particle present in a candle flame, or an incandescent liquid, such as a molten metal, emits a *continuous* spectrum of colour, that is, one in which all the wavelengths within its range are present. A colour spectrum emitted by a non-incandescent material, such as an excited gas, is quite different: it consists of a discontinuous selection of certain wavelengths only and is known as a *line* spectrum.

A flame is a gas in combustion—the consequence of a chemical reaction in which a gas combines with the oxygen in the air. In

[1] Sources which emit light, such as the sun, a flame or an electric lamp, are termed *direct* or *primary* sources, while non-emitting sources, such as pigments and other reflecting or transmitting surfaces, are termed *indirect* or *secondary*. This chapter summarises the direct sources available to the artist.

doing so, energy is released which is sensed as heat and light. The easiest way to colour the flame is to introduce into it a material containing a chemical *element*. The colour the flame subsequently appears is characteristic of the element (or elements) present in the burning process.

The introduction of the element sodium into a flame imbues it with a brilliant yellow appearance the quality and intensity of which can be observed by sprinkling common salt (sodium chloride) into a household gas flame. Using a spectroscope, this can be seen to be a mixture of the two predominant line emissions 589·0 and 589·6 nanometres. Burning potassium yields a distinctive purple flame which is composed of four principal spectral lines, two red (766·5 and 769·9 nanometres) and two violet (404·4 and 404·7 nanometres).

Other flame colourations include compounds of lithium or strontium for reds, calcium compounds for orange, compounds of tellurium or copper (except the copper halides[2]) for greens, and copper chloride for blue. The purest flame colouring is sodium yellow, which tends to mask all other colours when present. Each element or compound can be introduced into a gas flame as a water-solution spray just beneath the flame burner. Alternatively, and bearing in mind essential safety precautions, flame throwers can be used, fuelled by liquid petroleum gas (butane or propane) from a pressurised container; cylinders are currently available which give up to 30 hours continuous operation.

Other than introducing it into a flame, a material can be heated to incandescence by forcing a powerful electric current through it. For the artist this affords the advantage of a light source the intensity of which can be precisely controlled and regulated.

The most familiar incandescent electric light source is the tungsten filament lamp. This normally consists of a sealed globe of clear, etched or silica-coated glass fastened at its base into a metal bayonet or screw-cap socket. Inside the globe, glass stems support a fine coil-filament of tungsten thread which com-

[2] A copper halide is a compound of copper and a *halogen*. The halogens are the four related elements fluorine, chlorine, bromine and iodine.

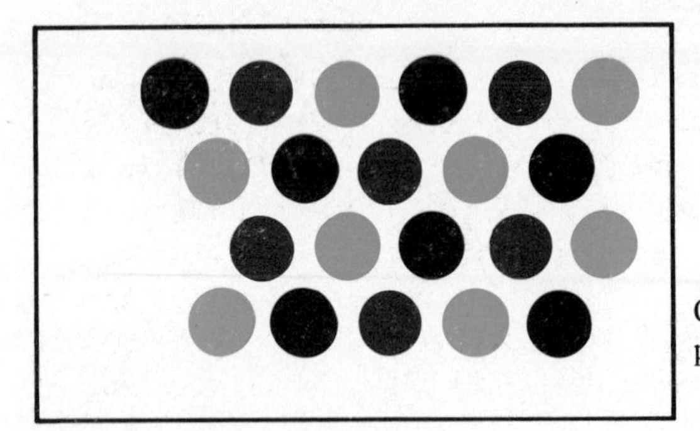

Colour 2.1 Television phosphor triads

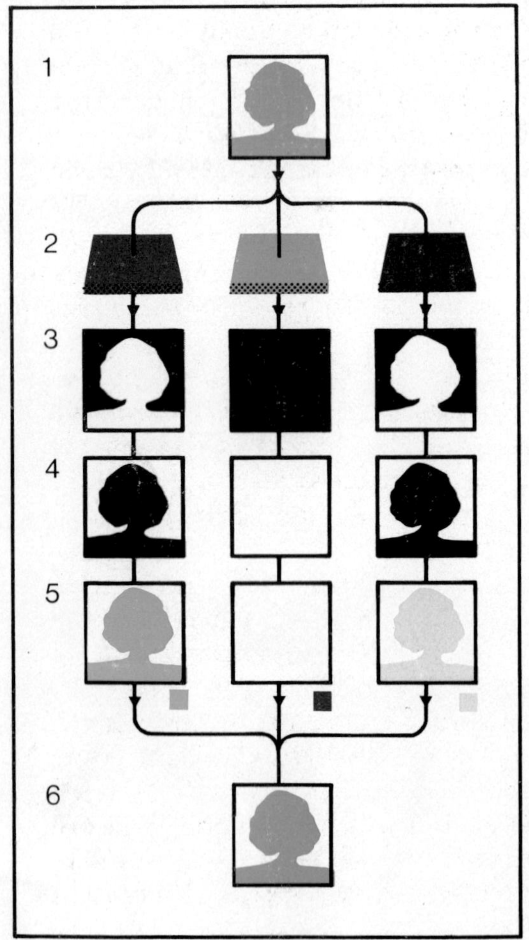

Colour 2.2 The principle of three-colour printing:

1 The object or copy is a green figure on a white ground

2 It is photographed in turn through orange-red, green and blue-violet filters

3 A set of black-and-white negatives is made

4 Corresponding black-and-white positives are made

5 The images are transferred to three printing plates and inked respectively with cyan, magenta and yellow inks

6 In the printed picture, the original object is reproduced by a combination of cyan and yellow inks

pletes an electrical circuit between two electrodes. When carrying an electric current, the tungsten wire is heated to incandescence and passes rapidly through red to white heat. Tungsten has a relatively high melting point (3,410 degrees Celsius) and achieves a much whiter light for instance than candlelight, since it can reach a much higher temperature. Compared with daylight, however, tungsten illumination is deficient in blue light and this must be compensated for (with filters) when matching colour to colour.

Oxygen is evacuated from the glass globe and replaced by an inert gas, commonly a mixture of argon and nitrogen, which retards the evaporation of the tungsten wire and secures for the lamp an economic life normally of 1,000 hours. By introducing a small quantity of a halogen gas, it is possible to set up a regenerative cycle which prevents black, evaporated tungsten depositing itself on the inner surface of the globe. This succeeds in both doubling the useful life of the lamp and stabilising its colour-rendering properties. Tungsten halogen lamps are therefore preferred for use in film projectors and photographic enlargers.

An incandescent filament lamp, with a platinum wire burner, was demonstrated as early as 1820. Not until 1879, however, did Edison evolve a commercially viable filament lamp close to its present form. Three years later, William Kurtz became one of the first photographers to boast 'portraits taken by electric light'; and in 1885 Edison collaborated with Louis Comfort Tiffany in the installation of electric lighting at the Lyceum Theater, New York. The filament lamp, the beam of which can be readily focused by lens, was subsequently adopted as the light source most highly suited to the projection of light beams.

In the *Lumia* light-projection system of Thomas Wilfred, reflective mirrors, lenses, filters and polished metal plates were employed to intercept light from filament lamps. In Wilfred's first *Clavilux* (constructed 1916–22), five manually operated sliding keys controlled the light intensity of a set of six lamps back-projecting beams on to a translucent white screen. It was given its first, silent performances at the Neighborhood Playhouse, New York, in 1922. Wilfred was to design over 160 *Lumia* compositions, three of which (dated 1941, 1955 and 1964 and programmed on automatic machines) are in the collection of the Museum of Modern Art, New York.

In 1928, Wilfred conceived a coloured-light spectacle on a particularly grand scale: a skyscraper surmounted by a 'Clavilux Silent Visual Carillon'. Ever-changing patterns of coloured light were to be projected on to the inside of a translucent glass dome and operated either manually or automatically. Something close to Wilfred's vision, and which could similarly have been conceived only in terms of electric light, was realised by Zdeněk Pešánek in the Czechoslovak Pavilion at the 1937 Paris Universal Exhibition. Pešánek installed a machine which controlled the operation of over 1,000 tungsten lamps. Projected patterns of light, utilising a full potential of almost 250 variations of colour, were recorded on interchangeable rolls of perforated paper fitted to a type of player-piano mechanism.

More recent examples of similar systems include those evolved in the United States by Jordan Belson and Christian Sidenius and in Europe by Frank J. Malina and Peter Sedgley. Belson, as visual director of the *Vortex Concerts* (1957–60) given at the Morrison Planetarium, San Francisco, programmed 50-minute shows integrating a 50-speaker sound system with as many as 70 simultaneous projections of light beams, films, transparencies and stroboscopes. Sidenius began work on his *Theater of Light* at Sandy Hook, Connecticut, in 1965, installing sophisticated equipment to control the multiple projection of lights, some incorporating Wilfred's *Lumia* principle, accompanied by music and choreography.

Malina and Sedgley have undertaken large-scale commissions but much of their work is small enough to be accomodated in the domestic interior. Both have developed methods of cyclic colour composition. In 1956 Malina devised his *Lumidyne* system of lamps housed in sealed boxes which illuminate from behind designs painted on replaceable translucent screens, some fixed while others rotate. Sedgley (since 1967) has affixed arrangements of dichroic filters[3] on to large circular panels (lit from the side by tungsten motorlamps) which slowly rotate, throwing part-reflected, part-transmitted beams of spectral light across the panel surface.

A highly incandescent light source alternative to the tungsten lamp is the carbon arc lamp, first demonstrated in 1810, which has no filament or globe. Instead, an electric current completes its circuit using air molecules as a bridge between two solid gas-carbon rods fixed very close to one another. The arc itself is not incandescent but the electrode tips of the rods are strongly heated and burned slowly away by the oxygen in the air. Craters in the tip of each rod become white hot and emit highly intense light (bluer than that of tungsten light) which approaches sunlight in its brightness.

The carbon arc lamp is most applicable where a highly intense light source is required and as such is used for searchlights and in colour film and television projectors. Carbon arclight was employed by Adrian Cornwell-Clyne in his *Colour Projector* of 1921. His machine consisted of a large spectroscope which dispersed spectral lights on to a small screen. Projection was controlled by a keyboard on which Cornwell-Clyne (known then as A. B. Klein) performed many London concerts accompanied by music, improvising from a range of over 250 combinations of coloured lights.

In types of non-incandescent arclight more efficient than the carbon arc lamp, the arc is enclosed in a cylindrical glass or quartz tube. When an electric current is passed or 'discharged' between the electrode ends of a tube containing a gas or metallic vapour at low pressure, a flow of electrons is released which excites the element present to emit its characteristic spectral wavelengths. In reference to this type of arc lamp it is necessary to distinguish between lamps containing gas at low pressure and those with gas at high pressure since increased

[3] A dichroic filter, or colour-selective beamsplitter, consists of a glass plate coated very thinly with a nonmetallic compound. Most types reflect the light of one colour while transmitting the light of its complementary.

pressure inside the tube broadens the emission spectra to wave-lengths beyond the lines normally associated with each element.

The most familiar 'discharge lamps' are the yellow low-pressure sodium vapour lamp and the bluish-white high-pressure mercury vapour (HPMV) lamp used in street lighting since 1933. Other low-pressure lamps include cadmium and zinc vapour lamps, which emit red light, the neon lamp, which emits orange-red light, the argon lamp, which emits blue-green light, and the carbon dioxide lamp, which emits near-white light. Among high-pressure lamps available, a new sodium lamp has been developed for street lighting, and the xenon lamp, emitting light similar in spectral composition to daylight, is used in compact film and transparency projectors; the HPMV lamp can be used with filters to obtain yellow, green, blue or violet light.

The first commercial discharge lamps appeared in 1904. The familiar neon lamp (patented in 1915) soon found its way into advertising display and more recently has become a popular art medium. The American artist Chryssa has used neon extensively in her work since 1962 and neon tubes have been 'collaged' with other media by, among others, Jasper Johns (1963–64) and Tom Wesselmann (1966).

Incandescence is always associated with high temperatures. Light emission by a material in which very little sensible heat is involved is known as *luminescence*. Almost all solid (or 'solid-state') materials capable of luminescence consist of a so-called 'host' crystal activated by an impurity, to which it usually owes its colour appearance. For example, the host crystal zinc sulphide appears yellow if its impurity is manganese, blue if silver, or green if bismuth or copper.

In electroluminescence, a form of luminescence occurring as a direct result of electrical activity, solid crystals are stimulated to emit light by forcing an electric current through them. One type of electroluminescent source utilises a 'sandwich' arrangement in which luminescent powder coats a reflective metal plate; an electrical circuit is made complete by placing a sheet of conducting glass over the metal and sealing the edges. Relatively little current is drawn and the whole flat panel (not just a filament or tube) is lit. The glass can be painted or stained, cut out and mounted in any position on floor, wall or ceiling. The intensity of the light emitted, which is generally low, is dependent on the magnitude of the current and the colour quality on the nature of the crystal(s) used.

Electroluminescent light is commonly used for the display lights of digital clocks, watches and pocket calculators. The emitting material is in the form of single crystals with wires attached individually, unlike the sandwich assembly in which millions of crystals are trapped between two electrode plates. Red, yellow and green are obtainable in the form of bars which make up the digits from seven segments. Efforts to obtain an efficient blue are in progress with the intention of having the three additive primaries available for mixing, since this type of electroluminescent light is in favour for the development of the thin colour television receiver.

A particular limitation in the use of the mercury vapour lamp is its absence of emission in the red region of the spectrum; this causes red objects to appear dark and the human complexion unflattering. A light source which imitates daylight more closely can be obtained by coating the inside of the mercury lamp with a luminescent powder, commonly an impure calcium halophosphate. The energy emitted by the mercury arc is rich in ultraviolet energy (185·0 and 253·7 nanometres) and the luminescence produced in the coating converts the invisible ultraviolet wavelengths into longer wavelengths within the visible range.

Luminescence may occur either during or after the absorption of energy at another wavelength. Emission which occurs only as long as the exciting input is being received is specified by the term *fluorescence*; emission which continues for some time after the energy input has ceased is said to exhibit the attribute of *phosphorescence*.

The coated fluorescent mercury lamp, introduced in 1938, is highly efficient in operation, a 40-watt fluorescent lamp giving as much illumination as a 150-watt tungsten lamp. It has a typical working life of upwards of 5,000 hours, which usually justifies the higher cost of its installation. Fluorescent lighting is ideally suited to area lighting but is not suitable for use with lenses to form spotlights; it is now used in almost all public buildings though incandescent tungsten light remains the more popular in the home.

In the common fluorescent lamp, the blend of luminescent crystals is selected to give white light, the standard commercial grades being 'warm white', 'cool white' and 'daylight', with *de luxe* versions. For special purposes, such as theatre or decorative lighting, fluorescent lamps can be obtained with coatings emitting red, green or blue light. The American artist Dan Flavin adopted the fluorescent lamp as his principal medium in 1961; using factory-made tubes in standard lengths and colours he has contrasted the various grades of lamp available, often subtly, as in *Daylight and Cool White* (1964: New York, Philip Johnson Collection), a simple assembly of four eight-foot tubes mounted together vertically.

Fluorescent pigments and dyestuffs appear more luminous than normal because they radiate more visible light than they receive. When sunlight falls on such a material, some of the invisible ultraviolet is re-emitted as energy of a longer wavelength, within the visible range. Practical applications of the phenomenon include the optical brightening agents used to treat papers and textiles and which are added to various commerical soap powders. Some of the ultraviolet energy absorbed by the agent is re-emitted as blue light which helps to cancel or neutralise the yellowness of the natural fibres.

Most luminous paints consist of colouring pigments intermixed with phosphorescent powder, such as an impure zinc sulphide. They appear to 'glow in the dark' because the energy they absorb in daylight is stored and released for a period of time after the energy input has ceased. Other paints are available which contain non-phosphorescent radioactive materials such as radium or thorium. The energy from these elements acts on other substances present in the paint and produces light emission which continues indefinitely, so that periodic stimulation by daylight is unnecessary.

Among the first painters to make use of fluorescent materials in their work were Frank Stella (1964) and Peter Sedgley (1966).

In 1960, at the Hughes Research Laboratories in Malibu, California, Theodore H. Maiman succeeded in producing a brief pulse of luminescent light the character of which was highly *coherent*, that is, all its waves were moving in the same phase or 'in step' with one another. This was the first LASER beam, an acronym for 'Light Amplification by Stimulated Emission of Radiation'.

An ordinary, incoherent light beam is the consequence of the independent or spontaneous emission of tiny flashes of light from single atoms, unrelated both in time and in direction of emission. Light energy can be made coherent by storing it inside a suitable luminescent material until a large number of atoms are ready to emit in unison. Excited by an emission wavelength of the element present in the material, the atoms are stimulated to reach their peak disturbance together and the stored energy is released in the form of a single, coherent and highly intense light wave.

The laser material used by Dr. Maiman was a rod of synthetic ruby crystal consisting of an aluminium oxide host activated by about 0·05 per cent of chromium oxide. Coiled around the rod, a xenon discharge lamp supplied the 200-microsecond flash of light sufficient to excite the chromium atoms in the ruby crystal. Most of the light is wasted but enough is of the right wavelength (694·3 nanometres) to be in tune with the waiting chromium atoms. In ordinary luminescence, each excited atom would drop back to its rest state independently and at random; but in the laser the stored light needs only a small boost at its tuned wavelength in order to stimulate a great many atoms to drop back together.

In subsequent designs, the ruby rod has silvered ends made accurately square to the axis of the rod, one having a slightly thinner coating than the other. Some of the light inside the rod reflects back and forth between these end-mirrors, the proportion increasing rapidly as more chromium atoms respond to the stimulus, until a powerful monochromatic light pulse bursts through the less-silvered mirror.

In 1961, a laser was operated which employed a mixture of helium and neon gases as its resonating medium. A radio-wavelength oscillator was tuned to a wavelength which excited the gas mixture whereupon laser activity was obtained. Unlike the crystal devices, which had given brief but powerful light pulses, the gas laser emitted a continuous, steady beam of light. Modern helium-and-neon (He–Ne) lasers are capable of emitting a variety of wavelengths in the red and yellow regions of the spectrum plus others in the infrared. Other suitable gases include argon, which emits in the green and blue regions, and krypton, which can be tuned to a selection of wavelengths througout the spectrum, though it is most efficient in the red; a helium-and-cadmium (He-Cd) laser is available which simultaneously emits five visible wavelengths which blend to give white light.

More recently, lasers have been operated which use luminescent crystals in liquid solution. The spectra of many of these 'dyes' are wide, rubidium dyhydrogen arsenate (RDA) for instance being capable of emission in the waveband 347 to 694 nanometres. Laser activity occurs over a specific waveband for each 'dyestuff' used, with tuning accomplished by altering the length of the cavity containing the solution.

Continuous-wave gas lasers, emitting steady beams of light, have proved the most useful to the artist; they exhibit the purest colours man has yet seen—which may be as much as a *million* times purer than light from an incoherent source. Two Americans who have successfully harnessed the laser for their own purposes are Rockne Krebs and Ivan Dryer.

Krebs (since 1968) has installed systems of mirrors and, on occasion, smoke-screens in order to display the remarkable optical properties of this unique light medium. *Canis Major* (1975–76), probably his most successful laser work to date, is housed in the Omni International complex of buildings in Atlanta, Georgia. The source is an argon laser, its single beam split by glass prism into a fan of four angled beams (two green and two blue) prior to interception by semitransparent mirrors located around the large interior courtyard of the complex. Krebs worked in liaison with the architect throughout the installation of this nocturnal laser piece, designed to complement his *Atlantis* (1973–76) which during the hours of daylight disperses the sun's rays through a precisely co-ordinated system of 1,000 glass prisms.

Ivan Dryer has presented laser shows in planetariums throughout America, Europe and Japan. In his *Laserium* concerts, first performed in 1973, a single krypton laser beam is split by prism into red, yellow, green and blue beams; an electronic control console permits the manual operation of computer tapes activating filters, oscillating and scanning mirrors and other prisms and lenses in the free improvisation of spectacular light patterns in synchronisation with a multichannel sound-speaker system. Dryer has envisaged the future of laser art both on a compact scale, small enough for the domestic interior, and on the vast scale of his proposed Dome Theater, which was to have housed the largest projection surface-area in the world.

CHAPTER 7

Additive Colour Reproduction

The most important application of the additive mixture of coloured lights is colour television. A rudimentary system of colour television was proposed as early as 1909 but not until 1928 did John Logie Baird demonstrate the first practical, electromechanical system. The Radio Corporation of America (RCA) subsequently evolved an all-electronic colour system, largely through the efforts of a team led by Vladimir Zworykin, and network colour television compatible with existing black-and-white broadcasting was inaugurated in the United States in 1953.

In 1962 Tom Wesselmann inserted a working television into a 'collage' of more traditional materials and television has since established itself as an important art medium. Nam June Paik began modifying broadcast images in 1963 and has estimated that by intercepting the electrical signals for standard television pictures he could create at least 500 image-distortions from one normal picture. Paik has since juxtaposed video images on tape and has also conceived a 'video synthesiser' to be used in conjunction with the standard home receiver to obtain colour distortions on the television screen which are controlled by the viewer.

The appearance of portable black-and-white videotape systems, made available by Sony in the United States and Japan in 1965, stimulated widespread interest in television and video art, and many experimental film-makers adopted the new medium. In 1969 Stan VanDerBeek became one such artist to be granted access to the professional equipment of WGBH-TV in Boston. This enabled him to extend his experience of computer film technology into the field of computer-programmed (rather than manually operated) video controlling and to experiment with the many electronic effects unique to television; these include mixing (tape editing), switching from one camera to another, and keying (inserting part of one picture into another).

Television is essentially a two-stage process in which visual and aural information about an original scene is transmitted as a composite electrical signal from camera to receiver.

The television camera is an optical device not unlike the human eye but, rather than arousing a photochemical response in the visual cells of the retina, light admitted through the camera aperture activates a mosaic of mutually isolated points, usually of a photoconductive material; small electrical charges are built up in the mosaic in accordance with the intensity of the incident light. In one of the simpler cameras, the RCA Vidicon tube (1952), a mosaic consisting of a thin layer of selenium or antimony trisulphide becomes electrically conductive on exposure to light, conductivity increasing as light intensity increases. A relatively inexpensive form of Vidicon, introduced for colour work by Philips in 1964, is the Plumbicon tube in which the photoconductive material is lead monoxide.

Red, green and blue filters cover the three apertures of the colour television camera. Each filter absorbs the light of its complementary colour and transmits an additive primary, enabling information about the light intensity of each separated component to be isolated on three mosaics ready for encoding in the camera. The mosaics are scanned by three beams of electrons (elementary electrical particles) directed to move across them from side to side and more slowly from top to bottom, like the eye reading a page. The points of the mosaic are encoded at a rate of some five million every second; short signals mark the end of each horizontal line and longer signals mark the end of each vertical scan or complete 'frame'. The filtered colour components are recorded simultaneously in accurate alignment and the signals combined for transmission to the television receiver.

In the receiver the conversion of electronic signals back to light is accomplished by a *cathode-ray tube* (CRT), known as the picture tube or kinescope. This consists of a large, evacuated glass container with a short cylindrical pipe at one end opening out to a flat viewing screen at the other. At the sealed end of the pipe are three electron guns, one responding to each colour signal; the current emitted by each gun is regulated by the intensity parts of the transmitted signal. The electron beams are sharply focused by magnet and deflected to scan the screen end of the cathode-ray tube by two electromagnets. These are controlled in turn by the line and frame markers in the transmitted signal, starting each line and frame of the televised picture at the right moment to give correct synchronisation with the original camera scan.

In the most common type of television receiver, the RCA colour tube (1949), the inner surface of the screen is coated with thousands of tiny, isolated points of phosphorescent materials each of which releases light in direct proportion to the current received from an electron beam. The 'phosphor' dots, each less than half a millimetre wide, are too fine for the eye to distinguish individually at normal viewing distance and blend optically to give the required colour impression.

The dots are arranged in sets of threes, known as *triads*, one of which emits orange-red light when stimulated by its appropriate electron beam, another emits green light and the third emits blue-violet light (*see* Colour Plate 2.1). Between the electron guns and the phosphor coating is a thin, perforated metal screen, the *shadow mask*, fixed about two centimetres from the phosphors. Its grid of perforations is arranged so that each hole lies precisely opposite a phosphor triad (Figure 7.1). This ensures that each electron beam falls on to the appropriately coloured dots and ultimately enables a faithful replica of the original scene scanned by the camera to be reproduced on the viewing screen by an additive synthesis of primary orange-red, green and blue-violet points of light. An alternative colour tube, proposed in 1957, employs sets of four vertical stripes (orange-red, green, blue-violet and blank) and eliminates some of the difficulties in precision manufacture and alignment of the phosphor triads and shadow mask.

In European systems, the television picture is composed of 625 horizontal lines of which about 90 per cent are used. There are roughly the same number of triads per line, giving over

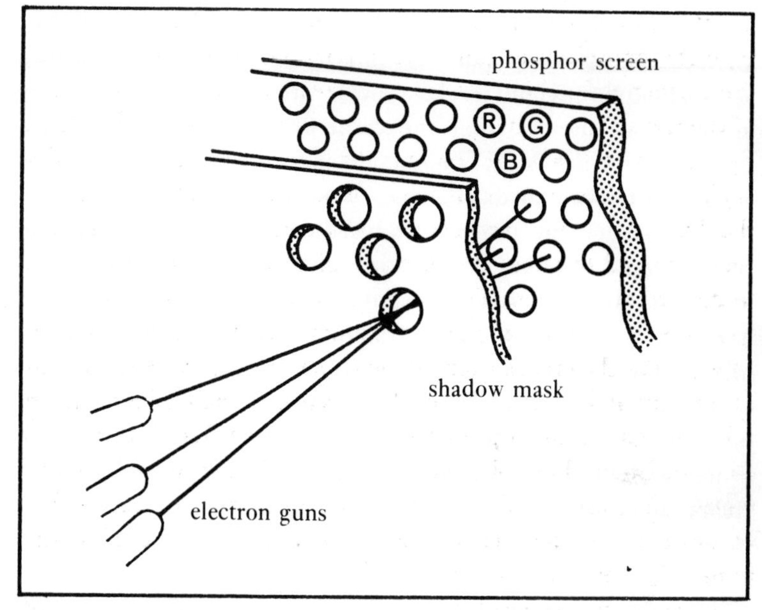

Figure 7.1 The RCA shadow mask tube (detail).

300,000 triads in all, each of which is bombarded by an electron beam approximately 25 times each second.[1] The eye cannot follow the movement of a scanning beam at 625 times 25 (equals 15,625) lines per seond, nor very easily the repetition of slightly differing images at 25 pictures per second, and the television image is preceived as continuous, with flicker only evident in the brightest picture areas.

There are several ways in which television signals can be transmitted from the camera to the receiver. The most common method is by broadcasting on the normal very high frequency (VHF) and ultra-high frequency (UHF) radio waves which are received by television antennae. Secondly, closed-circuit television (CCTV) links the camera directly to one or more monitors by wire or fiber-optic cable, similar to a telephone system. A school

[1] American and Japanese systems have 525 lines operated at 30 pictures per second.

or college may have a CCTV system in which a camera in a private television studio is linked by cable to monitors distributed throughout its classrooms. Alternatively the wire transmitting the signals from the camera can be plugged into a *videotape recorder* (VTR) and picture and sound recorded on a tape reel, tape cassette or disc.

The videotape recorder, introduced commercially by Ampex in 1956, records the transmitted information along the length of magnetic tape by a method similar to that employed in audio tape recording. Network programmes can be taped for future playback or live action recorded with a portable camera, to give instant or later replay. The videocassette, introduced commercially in 1973, has gained great popularity in the video market though it may in time be overshadowed by the videodisc, introduced two years later. The nickel cobalt videodisc, which plays on a turntable, permits a television programme to be stamped out and distributed cheaply.

Video tape, like audio tape, consists of a plastic (usually mylar) ribbon coated with magnetic particles, usually iron oxide; finer, chromium dioxide particles give a higher quality signal and are generally used in cassette and colour systems. The recording head of the recorder magnetises the tape coating in accordance with the signal it receives from the camera. When the tape is played back the recorded magnetic pattern induces a current which reproduces picture and sound on the television monitor.

There are many different kinds of video systems. They are most easily distinguishable by the width of the tape used; two-inch, one-inch, three-quarter-inch, half-inch and quarter-inch. As a general rule the wider the tape the more information and better image definition it is capable of recording. Two-inch tape is used almost exclusively in professional colour broadcasting. One-inch tape is used by business or educational institutions in which television demand is sufficient to support a private television studio; it permits precise editing and can accept several soundtracks. One-inch tape systems are not easily portable and all lightweight systems employ half-inch or, occasionally, quarter-inch tape. Battery-driven portable video systems, known as *portapaks*, consist of hand-held camera, tape recorder and monitor(s) plus microphone and earphones.

A system of large-screen colour television, for large-audience viewing, was demonstrated by Baird in 1935. Sophisticated modern television projectors include the Gretag *Eidophor* in which a powerful arc-lamp beam, intercepted by a colour-filter wheel and grating screens, projects a considerably enlarged picture on to a cinema-type screen. The upward size of this type of enlargement is limited to the image resolution possible with electron-beam systems. If each phosphor dot were not activated by an electron gun but received its current directly—in the manner currently used to make electroluminescent clock dials—then the size of the television picture would be relatively unrestricted.

In another system (proposed in 1969), orange-red, green and blue-violet laser beams are used in conjunction with signals from a standard television camera or videotape recorder. The beams are aimed at many-sided mirrors which are made to rotate in synchronisation with the transmitted picture signal. Light reflected from the mirrors is projected on to a screen some distance away, enlarging the television picture without loss of resolution, intensity or colour quality.

A method of image reproduction which relies wholly on the coherent character of laser light is holography, a form of three-dimensional photography. Its principle was stated in 1948 by Dennis Gabor; it then remained a mathematical curiosity until the early 1960s, when the development of the laser provided a light source sufficiently coherent and intense to make the original concept possible.[2]

Holography is a two-step process which consists of the recording and then the reconstruction of optical wavefronts,[3] both stages

[2] Emmett N. Leith and Juris Upatnieks, *Photography by Laser. Scientific American*, June 1965. Reprint 212.

[3] A wavefront is an imaginary surface linking points of equal phase in a wave propagated through a medium. For an unmodified laser beam its shape is a plane surface.

usually requiring a darkroom. Objects in space, illuminated by laser light, are recorded with standard photographic chemicals but without the assistance of a photographic camera.

Light reflected from the near or far surfaces of an object must travel different distances. A monochromatic laser beam, the wavelength and velocity of which are invariable in any one medium, can be used to 'measure' the distances traversed to the various parts of such an object, thereby describing its shape and contour. By intercepting the beam with a glass or plastic photographic plate, it is possible to record the *interference pattern* which occurs when the light waves reflected from the object collide with a second set of waves travelling directly from the same laser source.

When two trains of light waves are moving in the same phase, that is, when crest and crest coincide, they augment each other and constitute a single wave train which is additionally intense. Conversely, when the crests of one wave train coincide with the troughs of the other, the light waves cancel each other out and amplitude is zero. The resulting pattern of wave interference can be recorded photographically: spatial information is thereby encoded which can subsequently be used to obtain a faithful, three-dimensional image of the original object in space.

An interest in *trompe l'oeil* and stereo-optical illusion has attracted a number of artists to holography. They include Margaret Benyon, working in Britain 1968–76, and Salvador Dali, who collaborated with Lloyd Cross to make *Alice Cooper*, a cylindrical multiple-exposure hologram, in New York in 1973. Other than developing his own multiple-exposure (Multiplex) techniques (since 1969), Cross has pioneered hologram edition printing (1972), large-plate holography (1·2 × 14·5 metres, 1975) and motion-picture holography (1976).

To make a single-colour hologram, a laser beam is first split in two by directing it to pass through a semitransparent mirror. It is then necessary to modify each beam (using a beam-spreading lens) from one which is narrow to one which is cone-shaped. One of the beams, termed the *signal* or *object beam*, is aimed by mirror at the object to be recorded and the light

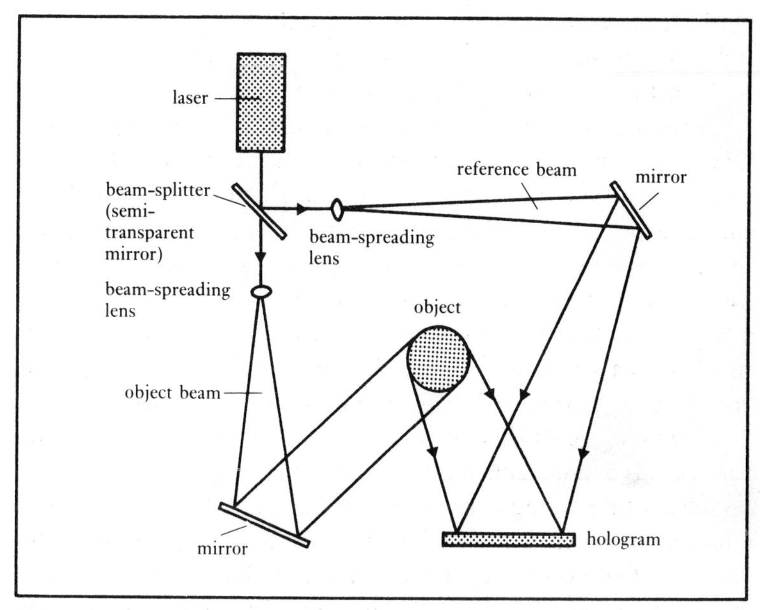

Figure 7.2 Recording a hologram.

reflected from the object is directed to fall on to an unexposed photographic plate. The second beam, called the *reference beam*, is aimed directly at the same plate (Figure 7.2). The two beams interfere as they are reunited on the surface of the plate; an exposure is taken, possibly lasting several seconds, and the unique pattern of interference is photographically recorded.

To reconstruct the optical wavefronts, the hologram, following ordinary photographic development, is illuminated by a reference beam of the same wavelength, direction and spread as the original beam. Wavefront reconstruction yields two sets of light waves which form equal angles to the plate and either of which can potentially be used to obtain the illusion: one set forms a *real image* on the side of the plate remote from the illuminating source and the other a *virtual image* on the same side as the source (Figure 7.3). The real image is characteristic of the reflection type

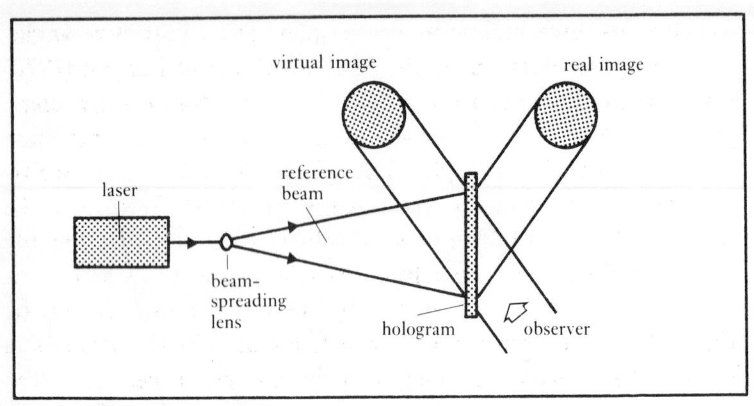

Figure 7.3 The principle of hologram reconstruction.

of hologram in which the image appears to float in space in front of the plate; the virtual image of the transmission-type hologram appears as though situated behind the frame of the plate.

If a hologram is lit by laser light of a different colour from its recording beam the object will appear larger or smaller according to whether the wavelength of the light is longer or shorter than that of the original. In 1969, a type of multicolour hologram was introduced which was viewable in ordinary white light, though its colours did not correspond to those of the original object. In this type of image, obtained by making a hologram of a hologram, transmitted white light is dispersed into broad horizontal bands of colour, hence its designation as a spectral or *rainbow* hologram.

It may be that full-colour holography, in which each colour appears in its appropriate position, will have been developed by the early 1980s. In theory, a full-colour reflection hologram may be obtained by exposing the object to be recorded in turn to orange-red, green and blue-violet laser beams and tilting the plate so that each exposure records its pattern on the plate at a slightly different angle. Reconstruction is then accomplished by illuminating the hologram with beams of the same primary wavelengths, each impinging on the plate at its respective angle.

An early development in the work of Margaret Benyon (1970) involved the triple exposure of an object in motion, using a single plate revolving about its vertical axis, and intended as a first step towards animated holography. The cylindrical (360-degree) Multiplex holograms made by Lloyd Cross (since 1973) require the transfer of over 1,000 individual film frames on to a flexible photosensitised plate; each frame is exposed as a stereo pair of narrow image strips, one viewed by the left eye and one by the right. The resulting rainbow-type hologram, in the form of a vertical loop made to rotate slowly around a central white light source, displays a motion-picture sequence of up to 45 seconds duration.

Cross has also made holographic 'movies' lasting several minutes but, as the flat conventional film frame is replaced by realistic depth-illusion, angle of view is critical and holograms of this type are at present limited to small-audience viewing.

CHAPTER 8

Pigment Sources

Nowadays, almost everything used, worn or consumed is coloured, and many thousands of dyes are in current commercial use. Some are suited particularly to the staining of natural fabrics, such as wool and cotton, while others are used to colour the many man-made fibres now available. Perishable foodstuffs dulled in processing are brightened with various chemical additives, and sauces, meats, cakes and conserves are generally heavily dyed. Other foods are bleached and recoloured, and artificial colourings play an important part in the preparation of confectionery, soft drinks and canned foods.

In industry, artificial pigments and dyestuffs have been employed to the virtual exclusion of natural colouring agents for a century or more. Housepaints, inks and dyes are made in quantity and the colour ranges offered by each manufacturer are displayed in specimen booklets or colour charts; the larger batches of paints are usually factory-mixed while others are prepared by machine in the retail store.

Until the nineteenth century, only a small number of artificial colouring agents were known. The great majority were of natural origin; their extraction from naturally occurring sources was difficult, expensive and time-consuming so that colourful clothes and tapestries became symbols of wealth and prestige. Hundreds of thousands of tiny murex shellfish had to be collected, and their pigment carefully extracted, so that a Roman emperor might wear his Imperial robe of Tyrian purple, while his subjects had to be content with uncoloured linen, hide and wool.

The three types of natural material available for use as colouring agents are the inorganic *earths*, the inorganic *minerals* and the organic *carbon compounds*. The earths are prepared by levigating (grinding and sifting in water) native clays high in metal content.

Of the 'iron reds', Red ochre is particularly rich in iron oxide and Sienna and Umber earths are rich in oxides of iron and manganese. Yellow ochre contains oxides of iron, aluminium and silicon, and Green earth contains silicates of iron, aluminium, magnesium and potassium.

The mineral pigments, obtained by pulverising natural ores, include Cinnabar (red mercuric sulphide), Realgar (orange arsenic disulphide), Orpiment (yellow arsenic trisulphide), Malachite (green copper carbonate) and Azurite (blue copper carbonate); Genuine ultramarine is prepared from lapis lazuli (a sodium aluminium silicate containing sulphur) and was once so expensive that it was the custom (as observed by Titian) for the patron of a painting to supply the blue pigment as well as the gold.

Glass is coloured in its manufacture by the addition of a wide variety of mineral compounds, each of which fuses with the raw materials of the glass to yield a characteristic colour. Oxides of cadmium, selenium or iron produce red glass, and those of cerium, titanium or vanadium give various yellows; oxides of nickel, chromium or tellurium give greens, and copper oxide gives either green or blue glass. Cobalt oxide is the blue colouring agent of the famous Venetian smaltini and Bristol Blue glassware of the seventeenth and eighteenth centuries. Manganese silicates yield purple, brown or black glass and they may be used to deepen many of the above colourations. Stannic (tin) oxide additives yield white or translucent 'opaline' glass.

The most important carbon compounds extracted from animal sources are natural Sepia, from the ink-sac of the cuttlefish, Crimson, the dried pigment of the kermes louse, and Carmine, the dried pigment of the cochineal insect, which very largely replaced natural Crimson in Europe after the sixteenth century.

Organic dyes derived from plant sources are extracted variously from roots, berries, flowerheads, barks and leaves. They include Madder, employed until fairly recently in the dyeing of red cloaks and uniforms, from 'dyer's root', the root of the madder plant. Other varieties of red are obtained from brazilwood, beetroot, cranberry, safflower ('dyer's thistle') and orchil ('dyer's moss');

orange dyes are obtained from the stigmas of the saffron, yellows from camomile and milkwort flowers, weld ('dyer's herb') or unripe buckthorn berries, and greens from ripe buckthorn berries or iris or ragweed flowers. Blue dyes obtained from the leaves of the woad plant (sometimes called 'dyer's weed'), have been popular since ancient times, often as a home-grown substitute for natural Indigo, a chemically similar dyestuff extracted from the plant of that name originally imported into Europe from India. Black dye from logwood chips is one of the few natural dyes still used commercially on a large scale.

As a rule, vegetable dyes are extracted by pounding or cutting up the colouring material; this is immersed in water, heated to just below boiling point and simmered until the colour has been transferred to the dye solution. For most natural dyestuffs to be effective the article to be coloured must first be saturated with a fixer or *mordant*, a mineral compound which 'bites' the fibres of the cloth in order to permit the insoluble adhesion of the colouring agent. A typical mordant consists of a 4:1 mixture of alum (potassium aluminium sulphate) and cream of tartar. The use of other metallic salts, as those containing chromium, copper, tin or iron, make it possible to obtain a considerable range of colours from a single dye source.

A process for the artificial extraction of Vermilion pigment, by roasting Cinnabar in a current of air, was known in Europe in the third century A.D. Methods for the preparation of a few other artificial pigments were known before then, including Verdigris ('Green of Greece', hydrated copper acetate), Egyptian blue (crystalline copper silicate) and Flake white (lead carbonate). The first modern man-made pigment was Prussian blue, a deep-blue compound of considerable staining power, synthesised at Diesbach's Berlin paint manufactory in 1704. Its process of manufacture, which was undisclosed for twenty years, is based on the chemical action of ferric sulphide on potassium ferrocyanide. The artificial formulation of other pigments followed, most importantly those of Cobalt blue by Thenard in 1802, Chrome yellow by Vauquelin in 1809, and French ultramarine by Guimet in 1826.

Before the nineteenth century, almost all dyestuffs were of natural origin, their chemical constitution unknown. The breakthrough came in 1856 when William Henry Perkin, an English chemistry student, accidentally discovered a vivid purple solution on mixing impure aniline (from coal tar) with potassium bichromate and alcohol. The commercial success of the resulting dye, known as Perkin's purple or Mauve, prompted the successful analysis of other natural dyestuffs, including those of Alizarin (from the madder plant) in 1868 and Indigo in 1880.

A wide range of natural organisms is preserved in coal deposits and it is possible by modifying the chemical constituents of coal to synthesise the wide variety of colouring agents which form the basis of the modern dyestuffs industry. Derivative paints and printing inks are the artificial *lakes*, which are prepared by depositing or precipitating a dye on to an inert, inorganic pigment (such as blanc fixe or alumina hydrate) which itself exhibits no colour response.

In 1936, the commercial introduction of copper phthalocyanine, from coal tar, made available an important new range of blue lakes highly fast to light and of exceptional staining power. Copper phthalocyanine, which closely resembles haemin, the colouring agent of chlorophyll, was introduced by Imperial Chemical Industries (ICI) as Monastral blue and, with chlorine added, Monastral green; most paint and ink manufacturers use their own proprietary names for the colours. Synthetic organic quinacridone dyestuffs, also from coal tar, have since their introduction in 1958 extended the range of rose and magenta lakes available, providing an artificial Alizarin substitute which is highly lightfast.

Artificial organic dyes have long since overcome the inconsistency of their notorious 'aniline period'. However, natural organic dyes are still popular and often preferred for use in traditional methods of hand-dyeing and batik and for the colouring of foodstuffs. Natural dyes are generally nontoxic; they are initially vivid in colour, blend together with age, and are well suited to the colouring of natural fabrics, such as silk and wool.

At the Paris Royal Manufactory of the Gobelins, natural dyes remained in use until 1911, when artificial colours were substituted. A revival in tapestry art, led by Jean Lurçat and inspired by a new freedom to interpret rather than merely reproduce a design, led to the reinstatement of natural organic dyes at the Gobelins in 1939.

A list of the most important pigments in current use is given in Appendix 3.

CHAPTER 9

Vehicles and Binders

Paints and inks consist of two substances: pigment and vehicle. The term 'pigment' is properly applied to the *insoluble* colourant powder which is dispersed and suspended in a transparent liquid, called the *medium* or *vehicle*. The colourant is often a natural or artificial oxide, and the fluid a natural or artificial oil, gum or resin, such as linseed oil, gum arabic solution or vinyl acetate. By contrast, dyestuff, the colouring agent of dye (almost always an organic compound) is *soluble* in its vehicle, which is usually water; it stains the material it colours and binds itself firmly to it.

Each surface coating possesses a specific percentage of vehicle in order to fulfil several requirements: it must support the pigment particles during storage in the tube or can; it must permit the coating to be easily applied and dry as a film which effectively binds the pigment particles together; and it must secure the permanent anchorage of the coating to the surface on which it is applied.

Once the paint or ink film has dried, it is the durability of the binding material which largely determines the useful life of the paint layer. In industry, paint films are generally not expected to last more than about ten years but as a rule an artist will want his colours to last much longer. Of the binders at present available, the synthetic plastic polymers exhibit the greatest strength in laboratory testing, though linoxyn (dried linseed oil) films have proved their durability over several centuries and dried egg-yolk (egg-tempera) films over as many millennia.

Linseed oil is a natural drying oil[1] extracted from the seed of the flax plant. It is soluble in distilled turpentine and has properties

[1] A drying oil is one which dries without the assistance of a metallic drier or siccative.

suited to a wide variety of artistic and industrial uses. As its volatile solvent evaporates, a film of linoxyn forms which holds the pigment fast; the oil absorbs oxygen slowly from the air and the initial oxidation is followed by slow chemical change which may continue indefinitely. The pigment is protected within a tough binding film which adheres tightly to the rough texture of the underlying material, traditionally a wooden panel or stretched linen or cotton canvas primed with a ground of white pigment.

A method of varnishing paintings with a solution of a resin in a drying oil was described as early as the eighth century but it appears that oil varnish was not adopted as a paint binder until some 500 years later. Popularised originally in the Netherlands in the fifteenth century, oil paint enabled the painter to preserve a vividness of colour and employ subtle effects of colour blending and overpainting (including glazing and scumbling) which were difficult or impossible with the alternative media then available. Brilliancy and gradation of colour are evident both in the *Portinari Triptych* (1476–78: Florence, Uffizi Gallery), by the Ghent painter Hugo van der Goes, and in the *Martyrdom of S. Sebastian* (1475: London, National Gallery) by the Florentine Antonio Pollaiuolo, a very early example of the use of the oil technique in Italy.

Drying oils, including those pressed from walnut kernels, hempseed, poppyseed and lavender (spike oil), in addition to linseed oil, were subsequently adopted throughout Europe as vehicles ideally suited to panel painting and they remained unchallenged until the large-scale availability of synthetic plastic vehicles (originally in the United States) in the mid-1950s.

In the earliest days of commercial printing in Europe, linseed oil was chosen as the ink vehicle most highly suited to letterpress (relief) and gravure (intaglio) printing methods. On the introduction of lithography (1798), existing oil vehicles were modified to meet the additional requirements of the lithographic process, such as the presence of water on the printing plate. In recent years, faster printing speeds have demanded new types of ink vehicle including the modification of the traditional natural oils with newer artificial resins.

A revival of interest has recently been shown in the ancient painting techniques of tempera and encaustic, both in extensive use prior to the introduction of oil paint. In the method of egg tempera, pigment is mixed or 'tempered' with a vehicle of fresh egg-yolk thinned with water. Egg-yolk is a natural emulsion containing water (about 50 per cent), non-drying oil and lecithin. As the water solvent evaporates, an oil film forms which holds the pigment fast. The yellow carotenoid dye of the yolk bleaches in sunlight and does not affect the colours of the dried painting. Ideally, tempera applied to a scraped gesso ground[2] gives a smooth paint film which dries quickly and without gloss.

As the tempera vehicle dries, the spaces occupied by water are replaced by air (an optically rarer medium). This causes a greater proportion of the light falling on the paint layer to be reflected, resulting in paling of colour coupled with loss of richness in the darker shades. The change may be remedied to some extent by coating the tempera picture with an oil varnish and by the early sixteenth century it had become customary to apply coloured oil glazes over a *lay in* in which the completed tonal scheme was worked out in tempera. Of the generation of artists who perfected egg-tempera painting, Sandro Botticelli adhered to the technique throughout his active career (c. 1465–1500); among modern painters, Andrew Wyeth has used egg-tempera extensively (since 1942).

Encaustic paint is the ancestor of wax crayon. In the original method, pigment mixed with a vehicle of glue and beeswax was applied to a dry surface in a hot, liquid state. The colours were then 'burned in' with a heated iron, giving a hardened paint film within which the bound pigment retained its brilliancy of colour and freshness of application. Such effects have been preserved for over fifteen centuries in the funerary portraits of Fayoum in

[2] A gesso painting surface consists of powdered calcium carbonate in a size binder.

Egypt. Encaustic was revived by Diego Rivera (1921) and Paul Klee (1933) and since 1954, Jasper Johns has executed many of his major works in a method combining encaustic with paper collage.

Wax crayon consists of pigment held in a stiff binder of fat or wax set into moulds to form short sticks. Among modern artists, Edvard Munch used the medium extensively through his long career (1885–1944). Wax crayon serves also as a wax resist in conjunction with water-soluble inks or dyes. In Javanese batik, now a popular wax-resist method, areas of cloth to remain undyed are covered with molten wax. The cloth is then immersed in a cold-water dye, which hardens the wax, after which the wax is melted with a heated iron and removed.

Oil crayon (or oil pastel) is a softer medium than wax crayon and suited to smearing, scraping or thinning with turpentine, by which methods colour mixture can be achieved. Dry pastel consists of pigment combined with a small proportion of non-greasy binder, usually gum arabic. Though rigid, pastel chalks are sufficiently fragile to disintegrate on contact with the rough 'tooth' of a suitable paper surface, after which the pigment is sprayed with a fixative (usually a thin solution of shellac) which surrounds the pigment particles and affords them some resistance to abrasion. Pastel was popularly used in eighteenth-century portraiture; among later artists, Edgar Degas was a prolific pastellist from the early 1870s until about 1908. Coloured pencils consist of pigment extended with clay and mixed in a gum-and-wax binder.

Aqueous or watercolour paint, in which solid pigment particles are suspended in a natural water-soluble binder, as a rule gum arabic, gelatin (size) or dextrin (starch gum), has been used extensively for several centuries. In the watercolour glaze technique, cakes of bound pigment are moistened with distilled water and transparent layers of colour applied with a soft brush to an absorbent white-paper surface. The white of the paper is preserved for highlights and superimposed coloured washes yield subtractive colour mixtures influenced little by the light-scattering effects associated with opaque paint systems; the method is illustrated in the watercolours of Paul Cézanne (1885–1906), and Paul Klee (1914–39).

The binding properties of gum arabic and gelatin were known in Ancient Egypt yet their adoption as paint vehicles was not widely adopted until early in the sixteenth century. Watercolour paint, a particularly portable medium, subsequently achieved immense popularity, initially in England in the eighteenth century; Turner was exclusively a watercolourist in the first years of his career (1792–96), and watercolour has proved highly popular with many modern painters, notably Emil Nolde (1907–45), Andrew Wyeth (since 1937) and Paul Jenkins (since 1954).

Watercolour paint is employed in both its transparent *aquarelle* and opaque *gouache* forms. Body colour and poster colour, being the cheaper grades of gouache, consist of watercolour paint extended with white paint to give it opacity and 'body'; applied to coloured paper they can be used to obtain colour highlights or to accent areas of local colour but they do not normally permit overpainting. Size colour (scenic colour) is water-soluble paint in which the vehicle is hot glue size, itself a solution of gelatin; it gives a quick-drying but brittle paint film, used almost exclusively for theatrical scene painting. Softbound distemper, consisting of pigment suspended in cold glue size, has been largely superseded by synthetic plastic paints but is still used occasionally as an inexpensive interior housepaint. A water-soluble vehicle, suited particularly to manuscript illumination (for which egg tempera is too oily) is glair, a preparation of egg-white whipped to a froth and allowed to stand to form a liquid which flows smoothly from the brush or pen.

At some stage of their formulation, vehicles and binders all possess the attribute of *plasticity*, that is, the capability of flowing or being moulded to a particular shape. Many plastic materials consist of chemical compounds which grow from small groups of short molecules, or monomers, into long chains, or polymers, by the process of polymerisation. Not all plastics are polymers though it is common for them to be so. Natural polymer plastics suitable for use as paint binders include gum arabic, gelatin and dextrin (all soluble in water), shellac (soluble in alcohol), casein (the binder of hardbound or washable distemper, soluble in ammonia) and cellulose, employed in a petroleum solvent in aerosol and spray paints.

The largest group of artificial or synthetic vehicles is the alkyd or glyptal resins (introduced in 1926) generally used in paint and printing ink systems in combination with natural drying oils. Alkyd resins are obtained by the polymerisation of certain alcohols with organic acids, one such reaction being that of glycerol with phthallic anhydride. Other important groups, some developed especially for serigraphy, include the coumarone-indene resins (from coal tar) and the phenolic resins, which include Bakelite, the first useful synthetic plastic (1909) polymerised from formaldehyde and phenol (carbolic acid).

Two synthetic polymers which make particularly durable and transparent pigment binders are the polyvinyl and acrylic resins, developed since the 1930s. In vinyl paint systems, vinyl acetate and vinyl chloride polymerise to form, respectively, polyvinyl acetate (PVA) and polyvinyl chloride (PVC); in acrylic systems, methyl methacrylate polymerises to form polymethyl methacrylate (PMMA). Two monomers are sometimes polymerised together to make a copolymer possessing properties which neither polymer would possess on its own. Vinyl chloride (a gas) is used in paint systems always to form a copolymer with vinyl acetate, though vinyl acetate (a liquid) is suitable for use alone.

Waterbound plastic or 'emulsion' paints are prepared by dispersing tiny drops of water with the solid pigment and monomer by high-speed agitation. This gives a vehicle in which either additional water or plastic extender can be employed as a diluent. Polymerisation occurs as the water evaporates and the fluid plastic coalesces into a tough, flexible and substantially water-resistant film within which the colouring pigment is firmly embedded.

Such paints are quick-drying (permitting rapid overpainting) and non-corrosive; they may be applied confidently to almost any surface which is dry, firm and free from dust and grease, suitable surfaces including primed or unprimed fabric, wood, wallboard, plaster or stone. Plastic emulsion paints can be applied by brushing, pouring, spraying or by spatula, enabling a body of paint to be built up into a thick paste or impasto. Additionally, foam-rubber rollers can be used to obtain speedily large areas of colour uninterrupted by hand pressure or brushmark. Diluted polymer staining methods, which in appearance resemble more a dyed than a painted finish, have been developed notably by Morris Louis (1954–62) and Paul Jenkins (since 1961). Opaque polymer painting techniques have been used by many contemporary painters, among them Barnett Newman (1967–70), Kenneth Noland (since 1962) and Frank Stella (since 1966).

Polymer plastic materials are of two classes: thermoplastics and setting plastics. Thermoplastic materials are hard and rigid at room temperature but soften and become plastic again when heated; they include the polyvinyl resins (PVA and PVC), the acrylic resins (PMMA, Plexiglas and Perspex), polyethylene (Polythene), polypropylene and polystyrene, which is employed in vacuum-forming sculpture techniques.

Setting plastics are of two types: thermosetting plastics (or thermosets) are plastic until heated, when a chemical reaction takes place which is irreversible; cold-setting plastics are plastic until undergoing a reaction at room temperature which similarly sets the material in its final form. The thermosets include the phenolic resins (Bakelite and Formica), urea-formaldehyde and most polyester (alkyd) resins; some polyester and polyether (epoxy) resins are cold-setting, though these usually set more quickly if heated.

Among artists who have chosen synthetic plastics as their principal media are Ronald Davis, John McCracken, Tom Wesselmann, Peter Jones and Maurice Agis; each has produced works which could neither have been conceived nor made in media other than those they employ. Davis has produced a series of 'plastic coloured' polyester wall panels (1966–72) reinforced with glass fibre and painted with pigment suspended in a polyester vehicle which fuses to become part of the panel itself. McCracken has manufactured a series of elegant glass-fibre-reinforced planks (since 1966) each impregnated with a single, vivid dye; and Wesselmann has created colourful panels in stitched PVC over foam rubber, issued in multiple editions (1971–73). Jones and Agis, in *Towards Colourspace Village* (1970–75) and *Colourscape* (since 1974), have collaborated to create large inflated structures of welded PVC sheeting which invite the spectator to move freely through a network of brightly coloured chambers of filtered daylight.

Glass is a manufactured thermoplastic consisting commonly of silicates of calcium and sodium, composed of silica, lime and potash or soda; traces of metal oxides are used to colour the glass or to modify unwanted colouration. Opacity, transparency or translucency (opalescence) are determined by the relative proportions of the raw ingredients, which form glass after three or four hours smelting at a temperature in excess of 1,500 degrees Celsius.

Coloured glass has been used as an art medium in Europe for over 2,000 years. In the technique of mural mosaic, small square *tesserae* of coloured glass have been used almost exclusively since the fourth century A.D. Sheets of cooled glass are cut into tesserae of about two centimetres square. This is large enough to prevent a dulling of colour by the plaster grouting which fills the crevices but small enough to permit fine detail in a fairly large format.

Glass mosaic was popularly employed in the vast decorative schemes of the early Byzantine churches, and splendid examples survive in Rome and Ravenna (fifth and sixth centuries) and in Sicily and Greece. The mural mosaics of the Church of Sta. Sophia in Istanbul, which were uncovered in 1922, exhibit designs incorporating some 2,300 different colours. Sensitive modern designs include those by Léger for the Church of Nôtre-Dame de Toute Grâce at Plateau d'Assy, Haute-Savoie (1946–49) and by Marc Chagall both at Nice University (1968) and in the Parliament Building in Jerusalem (1966–69).

Techniques which evolved from the development of glass mosaic are those of stained glass and enamelling. To stain the glass, metal oxides are either added to molten glass during smelting, to

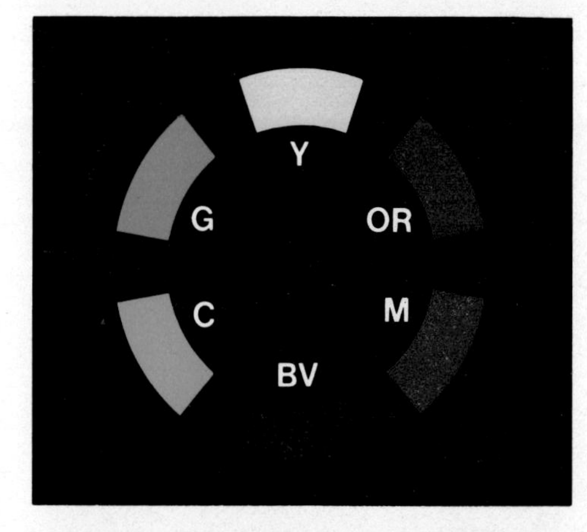

Colour 3.1
Runge's colour wheel

Colour 3.2
Ostwald's colour wheel

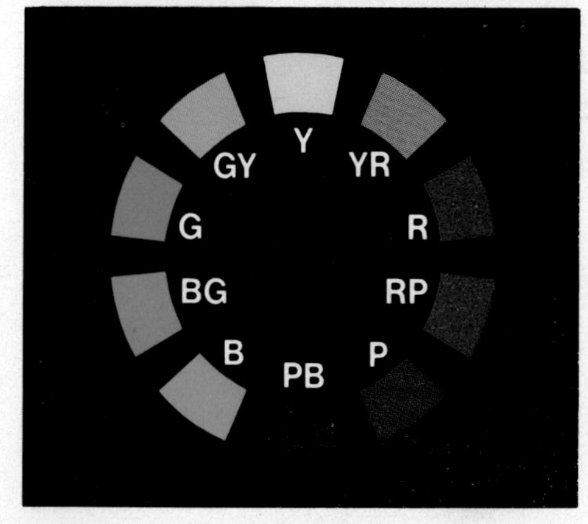

Colour 3.3
Munsell's colour wheel

obtain so-called *pot-metal* glass, or suspended in a gum solution and painted on to the glass before refiring, which fuses a *coated* stain or design to its surface. By 1100 the master glaziers of France were the leading exponents of the art and windows at Le Mans, Bourges and Chartres (which retains much of its original fenestration) testify to the great Gothic age of stained glass in Europe. Some of the tradition has survived: between 1919 and 1933 stained glass was taught at the Bauhaus (in turn by Itten, Klee and Albers) and impressive designs by Chagall include windows for the Gothic cathedral at Metz, Moselle, (1958–60) and the synagogue of the Hadassah Medical Centre in Jerusalem (1960–62).

Vitreous enamel consists of glass powder compounded so as to be able to fuse permanently to a metal surface, as that of copper, gold, silver, steel or aluminium, at a temperature of between 600 and 700 degrees Celsius. It is inlaid either into chiselled cavities, known as *champlevé* work, or between soldered partitions, known as *cloisonné* work. Unrivalled enamelwork by the Byzantine goldsmiths is exhibited in the Golden High Altar of the Cathedral of S. Marco in Venice (tenth to twelth centuries). Georges Rouault, late in his career, designed a number of enamels (1949–58) and Roy Lichtenstein has issued series of 'painted enamel' pictures in multiple editions on steel (1964–66).

In an alternative form, glass enamel known as *slip* is used as a ceramic glaze. Slip is generally applied to pottery by dipping an article to be coloured into a slurry of enamel powder (frit), chemical salts, fine clay and water; designs may be painted on to its surface or scratched (*sgraffito*) through a slip coating prior to refiring. Ceramic glazes were in use 5,000 years ago in Egypt and perfected in China during the Ming Dynasty (1368–1643). Among modern European artists who have applied their originality to the method are Paul Gauguin (1886–95), Chagall (1950–57) and the potter Bernard Leach (active 1911–74).

Slip is fused to slabs of fired clay to make ceramic tiles for use in large-scale mural work. The medium was employed extensively by Gaudí in Barcelona (1887–1914) and by Rivera in

Mexico City (1951–55). Léger produced a remarkable series of polychrome ceramic pictures (1950–55) and a Léger design of 1954 was interpreted in ceramic tile for the façade of the Fernand Léger Museum at Biot, Côte d'Azur.

CHAPTER 10

Subtractive Colour Reproduction

Subtractive colour mixing has two principal commercial applications, one in colour printing, the other in colour photography and film.

Though a method of colour photography 'composed of three pictures superimposed, the one red, the second yellow, and the third blue', was outlined in 1862 by Louis Ducos du Hauron, the unavailability of suitable filtering dyes did not permit him to produce his first successful set of colour separations until some seven years later.[1] By the end of the century the three-colour process of photomechanical picture reproduction had become widespread, following the appearance of the first high-quality photoengraving, made by William Kurtz and printed in a Boston journal in 1893.

In printing, subtractive mixing is suited to the reproduction of pictures in full colour, as those familiarly seen in glossy magazines, posters, picture books and packagings. In each case a paper or plastic base acts as a white reflecting ground on which successive layers of transparent inks or dyes are deposited. Each layer selectively absorbs part of the incident illuminating light, which is usually white; the rest of the light is ultimately reflected to the eye of the viewer.

In order to reproduce a full-colour copy or facsimile of an original scene, it is first converted into three photographs each of which records one third of the total colour information. The separation of the image into its colour components is accomplished by placing three differently coloured filters over the aperture of the camera and taking a black-and-white photograph through each in

[1] Maxwell had demonstrated the principle of three-colour photography by additive mixing, adopted in several early systems, in 1861, but all modern systems employ the subtractive principle.

turn. Each filter transmits one third of the picture information while absorbing the other two thirds: the orange-red (red) filter absorbs the cyan-blue component, the green the magenta-red, and the blue-violet (blue) the yellow component.[2] A set of three negative transparencies is thereby obtained which serves as the basis for the recomposition of the picture in full colour.

Taking as an example an object which appears *green* in daylight (ideally reflecting wavelengths in the middle of the spectrum but none in the red and blue regions), this is photographed in turn through the red, green and blue filters. Of the three, only the green filter transmits green light, which causes opacity or *density* in the green-filter negative. Neither the red nor the blue filter transmits green light so that opacity is not produced in the red- nor the blue-filter negatives. At this preliminary stage, in which the negatives are developed photographically to a suitable degree of contrast, the green-filter negative is *opaque*, the red-filter negative is *transparent* and the blue-filter negative is *transparent* (*see* Colour Plate 2.2).

In the next stage, a photographic positive is made from each negative. This is done by transmitting light through each in contact with a second unexposed photographic plate. As a positive has optical density where a negative has none, the situation is reversed, thus: the positive made from the green-filter negative is *transparent*, the positive from the red-filter negative is *opaque* and the positive from the blue-filter negative is *opaque*.

In the final stage the positives are transferred on to a set of three printing plates or blocks prepared with photosensitive chemicals.[3] The positive image separated from the original scene by the green filter forms the *magenta* component of the printed image, the

[2] Alternatively, using two colour-selective (dichroic) beamsplitters, the light from the original scene or copy is directed to pass through a blue-reflecting filter, which isolates the blue component and transmits the red and green, and then through a red-reflecting filter, which separates the red from the green.

[3] Materials which undergo a physical or chemical change as a result of exposure to light include ammonium or potassium dichromates, used in printing to transfer the photographic image to the printing surface, and those based on a silver halide (usually silver bromide or silver iodide) which are used in conventional photography.

image separated by the red filter forms the *cyan* component and the image separated by the blue filter forms thc *yellow* component. The set of three plates or blocks are then printed in succession, one on top of the other and in accurate alignment. In the instance of the photographed green object, the magenta component is found to contribute nothing to the final image; but both the cyan and yellow components do contribute and, when combined as transparent ink layers on white paper, they mix to yield green, thereby reproducing the green of the original object.

Ideally the band of wavelengths passed by each filter dye should correspond exactly to the waveband absorbed by the inks employed in the printed reproduction; that is, a filter dye and its corelative ink dye should be precisely complementary in colour. In practice, however, such close correspondence is unattainable; owing to the impurity of the light transmitted particularly by magenta and cyan colourants currently available, the copying of a full-colour scene by subtractive mixture is achieved only imperfectly and the range of colours imitated by intermixing the pigment primaries is more restricted than one would prefer.

A consequent defect of the three-primary process is that in practice it often does not yield a sufficiently dense black (a fault more serious in colour printing than in colour photography). In order to overcome this problem, a fourth *black* image-component is customarily added to the three coloured components. On the basis of complementary colour pairing, the black-image filter should be 'white', that is, no filter would be necessary, but a yellow filter is usually used in order to normalise the exaggerated blue sensitivity of standard photographic film.

Though the supplementary black image is a refinement (and in theory not an essential part of the full-colour process) it is the black picture-component which largely determines the light-to-dark value or *tone* of the printed image, while the combination of primary inks determine its colour quality. The black adds definition and contrast to the colour reproduction and is often needed to dull the appearance of the primary colours, which might otherwise appear too vivid.

In the gravure printing method, variations in tone and colour quality can be obtained by varying the thickness of the printed ink film. This requires the preparation of a set of *continuous-tone* separation photographs in which the gradation from highlight to shadow is smooth and unbroken. In other printing methods, principally letterpress, lithography and serigraphy (representing respectively relief, planographic and stencil printing methods), it is not possible to apply different thicknesses of ink to the paper for the purpose of obtaining variations of tone and colour; the ink layer lies in a single plane across the printing surface and it is a matter of uniform inking or no inking.

The problem is solved by dividing a photograph into a mosaic of separate dots which are too small to be detected individually at normal viewing distance. This method, called the *half-tone* process, is accomplished by rephotographing each colour-separated image through two glass plates incised with closely ruled parallel lines and bonded together at right-angles. The half-tone images are then transferred to the printing plates or blocks and the resulting effect, shown enlarged in Figure 10.1, can be seen in a magazine colour picture.

The dots which make up the picture vary in size according to the apparent tone of the original photograph. From the appropriate viewing distance, which may be several metres for a street poster, the dots blend optically and give a convincing illusion that tone and colour gradations are continuous. In order to avoid the type of optical interference known as periodic or moiré patterning, the grids of dots are printed diagonally across one another at angles of 15 or 30 degrees apart. In this arrangement the larger half-tone dots are superimposed and give straightforward subtractive mixtures. The smaller dots however are intended to lie in juxtaposition only, so that a substantial part of the overall impression results from optical colour mixing which is *apparently* additive in principle.

The development of electronic half-tone engraving machines (largely since the 1930s) has enabled half-tone colour separation plates or blocks to be produced automatically. The principle of such

Figure 10.1 Enlarged half-tone newspaper picture.

scanners is not dissimilar to that of the television camera but scanning is achieved usually with a thin beam of light (from a xenon lamp, or an argon laser in the newer machines, since 1972). Sharper image definition is also required in the printed picture so that scanning is relatively slow and definition of the order of 65 to 200 lines per inch (24 to 80 lines per centimetre).

Richard Hamilton's editioned serigraph *Bathers* (1967; 70 × 95 centimetres) consists of a colour-separated print modified through several stages of enlargement from part of a 35-millimetre colour transparency. His earlier *Whitley Bay* (1965; cellulose paint on photograph) exhibits the deterioration of picture detail which occurs when a photographic half-tone is highly magnified. In paintings and prints by Roy Lichtenstein (since 1962) the half-tone process is parodied, many of his images incorporating grids of Ben Day dots stencilled in red, blue and yellow.

Black-and-white photographic film consists of a cellulose acetate support carrying a single layer of light-sensitive chemicals, usually an 'emulsion' of silver halide crystals suspended in gelatin. Colour photographic film, known as *integral tripack* film, consists of a single transparent support carrying three such layers of crystals. Photographic emulsions are ordinarily sensitive only to the blue region of the spectrum and such an emulsion constitutes the top layer of the tripack. Below it a yellow filter prevents the transmission of blue light to the lower layers, one of which is sensitised to red light and the other to green. When the film in the camera is exposed to light reflected from the scene, some of the metallic silver separates from the silver halide, recording a colour-separated image (as yet invisible) in each chemical layer.

Colour transparency films (which are those intended to be viewed by transmitted light) are produced from a type of tripack known as *reversal-colour* film. After exposure, the film is processed in two stages. The first is the standard chemical development in which the invisible or latent image, consisting of metallic silver grains, is amplified many millions of times and made visible. In the second stage the film is exposed again (chemically or with light) causing a 'reversal' of the negative colour image into a positive one. The film is developed a second time; chemical reduction then occurs in which the halide in each layer is further separated from the metallic silver. As the developing solution becomes oxidised it reacts with chemical *couplers* contained in each emulsion layer. The couplers combine with the oxidised compounds to produce insoluble dyes which, in reversal-colour film, correspond to those of each subtractive primary.

The photograph is made permanent by *fixing*, a process in which the silver halide crystals unaffected by the developing solution are converted into a water-soluble silver salt which is then washed away. In the resulting positive transparency print the three dye-images combine by subtractive mixture and, when illuminated by a beam of white light focused from a suitable projector lamp, the superimposed magenta, cyan and yellow components contribute to render a faithful copy of the original scene in full colour.

Using special processing chemicals it is possible to make a positive photographic print on paper from a positive colour transparency. Opaque paper prints (or positive reflection prints) are however usually printed from *negative-colour* film, in which dyes are formed during the initial development. After exposure in the camera to light from an original scene, each chemical layer of the tripack is developed into a black-and-white picture component. The oxidised developer then reacts with the couplers to form the appropriately coloured dyes. In this case, however, once the unwanted silver halide has been washed away the film carries a *negative* record of the photographed scene, that is, one in which each colour of the original is replaced by its complementary colour.

To obtain a print, white light from an enlarger lamp (in the darkroom) is passed through the processed negative and directed to fall on to a sheet of photosensitised paper also coated with a three-part emulsion layer. The colours of the negative are reversed at this stage so that in the resulting positive print each colour of the original will appear in its appropriate position.

Reversal-colour film is used by most photographers to obtain readily saleable transparencies intended for reproduction in books, magazines, albums and so on. Negative-colour film is usually reserved for experimental and exhibition work; unlike transparency processing, printing from negative-colour film allows considerable opportunity for the manipulation and adjustment of the exposed photographic picture at a creative level.

Two photographers who advocate such experimentation, and whose work has been widely published and exhibited, are Fred Burrell and Robert D. Routh. Burrell's methods (since 1954) include those of masking, in which particular sections of the image are shielded from the light of the enlarger lamp during printing, toning, in which colour is added to a developed black-and-white print by single or successive submersion in various dye solutions, and posterisation, a process involving the sandwiching of high-contrast positives of the same image, exposing each through filtered light to obtain sharply defined areas of flat colour (imitating those of poster paint). Routh (since 1960) has additionally utilised the technique of solarisation, in which either the negative or the print is re-exposed to varying amounts of light during development, resulting in partial or complete reversal of colour and tone in the finished photograph.

A photographic camera consists essentially of a lightproof box in which light reflected from an original scene is admitted through a small aperture at one end and focused by lens on to a film frame at the other. The motion-picture camera employs the same optical system as the still camera but has additionally to accommodate a sequence of high-speed exposures at regular intervals along a continuous strip of film. Its shutter is therefore mechanically synchronised with the passage of film past the aperture at exposure-speeds normally of 16 frames per second for silent film or 24 frames per second for sound film.

The first successful integral tripack film was introduced under the trade name Kodachrome in 1935, originally in the 16-millimetre gauge for the amateur film-maker; it has subsequently become highly popular in the 8-millimetre (Kodak Super-8) gauge. Though the introduction of faithful colour reproduction meant an important new dimension in the development of cinema art, few feature-film directors have exploited colour in much more than a matter-of-fact way. Exceptions include Michelangelo Antonioni in *The Red Desert* (1964; 120 minutes), his first colour feature, in which exaggeration and extinction of colour serve to emphasise contrasts of mood, and the animated films of Walt Disney, notably *Fantasia* (1940; 138 minutes), and George Dunning, in *Yellow Submarine* (1968; 87 minutes), both of which employ a wide variety of graphical and film techniques.

Films which exhibit an experimental and innovative approach to colour cinematography are more often short, personal and non-commercial, as those of Len Lye (1932–52), Jordan Belson (since 1952) and Robert Breer (since 1966).

In *A Colour Box* (1935; 5 minutes), Lye pioneered the technique of 'cinepainting' in which cellulose-based dyes are applied directly to raw film stock (thereby dispensing with the film camera). Vividly coloured patterns are synchronised with rhythms on the sound track, an approach also adopted by Belson in his *Phenomena* (1965; 6 minutes), which combines electronic music with filmed sequences of distorted television pictures and cloudlike patterns of projected light. Breer's lively animations, some photographed from flip-cards, explore the visual response to discontinuous sequences of images which can induce hallucinations of colour and form in the perception of the viewer.

PART III

CHAPTER 11

The Measurement of Lights

Before the nineteenth century the visual artist needed to concern himself with only two sources of artificial light: the candle and the oil burner. Following the introduction first of the gas lamp (1798) and then a bewildering array of electric lamps, the intensity and colour-rendering properties of different illuminating sources have become essential considerations. The present-day manufacture of lamps for specific purposes, such as studio or gallery lighting, display lighting or photography, has called for a precise, objective and unambiguous system of light measurement.

A most remarkable feature of human vision is its ability to adapt to widely different levels of illumination. So flexible is this faculty that variations in light intensity of a ratio of as much as 100 to 1 may be scarcely noticed if the change is not too sudden. However it is this feature of adaptation, in which visual sensitivity effectively diminishes in bright light and increases in dim light, coupled with the subjective and inconsistent character of individual colour vision, that makes the eye a quite unsuitable instrument with which to establish a reliable standard of light measurement.

The branch of science which deals with the quantitative measurement of light is photometry; and instruments which permit the comparison of the intensity of light sources are called photometers.

The first effective photometer was constructed early in the eighteenth century by Pierre Bouguer. Its principle still forms the basis of modern devices in which two light sources are directed to illuminate two identically white surfaces. One source is a standard source and the other a test source. The standard source and the two surfaces are fixed in position while the test source is free to be moved closer or farther away from the surface it illuminates until both surfaces are judged equally bright.

A beam of light travelling outwards from a point on the

surface of a light source, whether flame or electrical, moves in a straight path and spreads out in both transverse directions. At twice the distance from the source the amount of light flowing through the beam is spread by one factor of 2 horizontally across the beam and another factor of 2 for the vertical spread. In other words, a cone-shaped beam of light flows outwards from the point source and its intensity is found to vary inversely with the *square* of the unit distance from the source.

At double the distance from the source, light intensity is quartered. At three or four times the distance the light is spread over 3 times 3, or 4 times 4, times the area of the end-section of the cone at 1 unit of distance from the source. This rule, called the *inverse square law*, was proposed by Kepler in connection with the measurement of light as early as 1604 and it remains today a fundamental photometric principle. Using a photometer, the law enables the intensity of the test source to be determined relative to that of the standard source; but this is only a comparison. The objective measurement of light cannot be accomplished until a *standard unit* of light flow has first been established.

Early attempts to obtain a practical standard involved the use of 'standard candles' made according to a defined formula. The candles did not turn out to be very reproducible and in 1800 the more controllable oil burner was adopted, followed by the gas burner and then the electric filament lamp. None of these illuminants was found to possess the desired permanency and since 1948 the modern standard has consisted of a beam of light from an aperture in an electrically heated furnace kept at the temperature of freezing platinum. Pure platinum always solidifies at precisely the same temperature (2,042 degrees Celsius); and the amount of light emitted per unit area of the aperture per second by the furnace depends on temperature only, provided that the aperture is only a small part of the total enclosed surface. An aperture of area 1 square centimetre in a furnace operated at this temperature always emits a specific amount of light per second; this has been defined as 60 *candelas*.

A source of 1 candela intensity emits a specific amount of

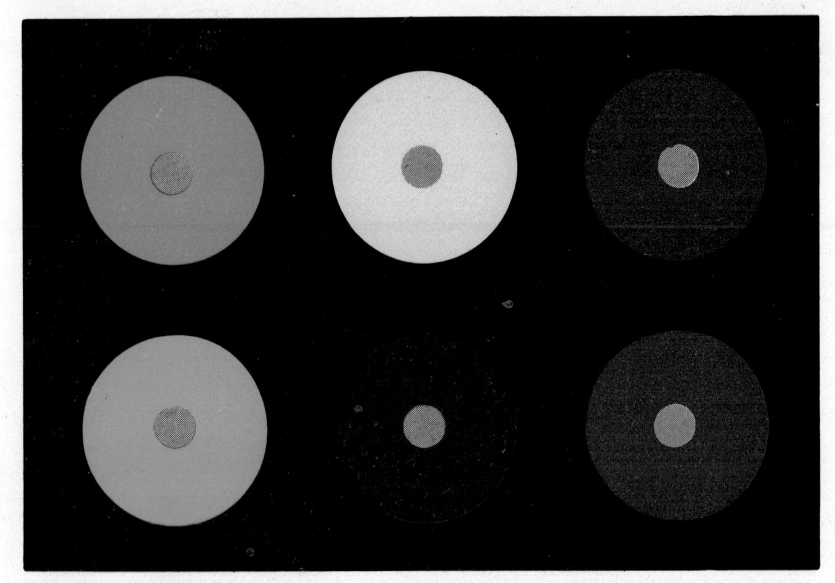

Colour 4.1 Simultaneous colour contrast. Complementary colours are induced in each grey centre

Colour 4.2 Successive colour contrast. Look for ten seconds at the coloured figure, then look at the white figure

light into 1 unit of cone per second; in the International System (SI[1]) of units, this amount of light has been given the name 1 *lumen*. A unit of cone, or unit of *solid angle*, as an alternative name,[2] is one where the area of the open end of the cone is equal to the square of the distance from its apex (Figure

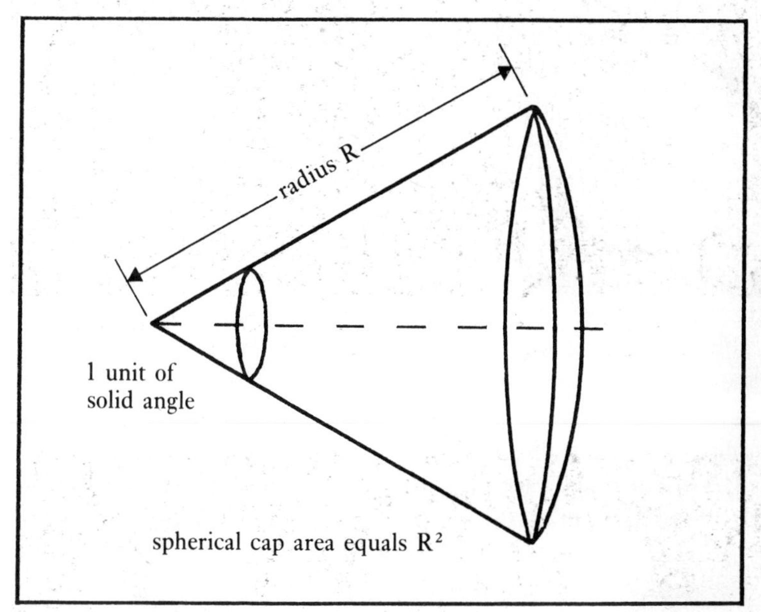

Figure 11.1 A solid angle or cone.

11.1). If such a source were positioned at the centre of a sphere, the *luminous flux*, being the total amount of light flowing in all directions from the source per second, would total 4π (equals 12·56) lumens. This is because the total solid angle of a sphere is 4π : its surface area is $4\pi R^2$ and its radius squared is R^2, so that the ratio of area to radius squared is 4π.

[1] *Système Internationale*, an internationally agreed system of units derived from the metre, kilogramme and second.

[2] The SI unit of solid angle is 1 *steradian*.

It is a matter of definition that 60 lumens of light are flowing per second through each unit of solid angle from a platinum-furnace aperture of 1 square centimetre. The area of a *light meter*,[3] placed at known distance from the furnace, will occupy part of 1 unit of solid angle, which can be calculated. In an example, a light meter of area 10 centimetres by 10 centimetres is set 10 metres away from a platinum furnace. The area of the light meter (0·01 of a square metre) divided by the square of the distance from the aperture is 0·001 of a unit of solid angle; the area of the meter is therefore intercepting 0·06 of a lumen of light. If it is portable, the light meter can be taken away and used elsewhere. Whenever it shows the original reading it is receiving 0·06 of a lumen of light. As the area of the meter is known to be 0·01 of a square metre, the illumination falling on it must then equal 6 lumens per square metre.

If the meter has been proved to read half as much for half the light, twice as much for twice the light, and so on, it can be used to measure illumination at other levels; and this in essence is what the reading indicates on an exposure meter used by a photographer or illumination engineer.

The lumen has been referred to as the unit of luminous flux or 'light flow' in all directions from a source per second. Of the SI units derived from the lumen, one has been mentioned but not yet named: the *illumination* falling on the surface of the light meter was given in terms of *lumens per square metre* and 1 lumen per square metre is known as 1 *lux*. One candela was given as a unit of luminous *intensity* and corresponds to a light source emitting at a rate of 1 lumen per unit of solid angle. Finally, the number of lumens conveyed to the eye from unit area of a source or surface per second flowing through a cone of 1 unit of solid angle is known as the *luminance* of that source or surface; the SI unit of luminance is 1 *nit*.

Luminance is related directly to optical impression, indicating how much light is conveyed from the surface of an object to the

[3] A light meter is an instrument containing a photoelectric cell which converts light into an electric current which can be measured.

eye of the viewer. For example, if a chessboard is placed under an electric spotlamp, while both its black and white areas are receiving the same amount of illumination, a black square will exhibit considerably less luminance than a white square. Though the part of the illuminating light reflected by either area can be defined in terms of lumens per unit area, the nit is usually used to indicate brightness in a specific direction.

If the illumination falling on a surface is known, its luminance can be calculated if the *luminance factor* or (for reflecting surfaces) *reflectance* of the material is also known.[4] Reflectance can be ascertained by comparing the proportion of incident light reflected respectively by two surfaces under identical conditions of illumination and viewing. If one surface is perfectly white,[5] reflecting 100 per cent of its incident light, the luminance of the test surface divided by the luminance of the perfectly white surface will indicate the reflectance of the test surface as a percentage ratio. In practice the reflectance of opaque surfaces which look white is found to be in excess of 80 per cent, while that of black surfaces is generally below 10 per cent.

Photometry achieves the measurement of only one parameter of the three needed to determine fully an objective description of a coloured stimulus. While luminance expresses the physical quantity of light conveyed per second from the stimulus to the eye, it gives no indication of its physical quality. The two other concepts required are *dominant wavelength* and *purity*, the colorimetric dimensions which together constitute the physical colour quality or *chromaticity* of a stimulus.

In subjective terms, a colour can look red, orange, yellow, green, blue or violet and one colour more or less colourful than another. Objectively, the colour can be reasonably well described by its corresponding spectral wavelength and its colourfulness by its purity. The concept of 'dominant wavelength' is

[4] The corresponding term for transparent or translucent surfaces is *transmittance*.

[5] A magnesium oxide coating, newly deposited on silver or aluminium to a depth of 1 millimetre, is usually adopted as a perfectly white surface.

applied to coloured stimuli other than actual spectral lights and is an estimate of the spectral wavelength to which the stimulus most closely corresponds. Purity is a measurement of the degree to which coloured light departs from monochromatic light (purity value 1) and approaches white light, a mixture of all spectral wavelengths, which exhibits zero purity.

Three-colour colorimetry is based on the assumption that any colour can be imitated in appearance by the additive mixture of three amounts of primary lights. If the spectral composition of the chosen primary lights were known then it would be possible to encode the sample colour by simply measuring and noting the intensity of each component primary. Instruments designed to enable such a match to be made are called colorimeters; and the three primaries needed to make such a match are usually called the *matching stimuli*.

By proposing that the matching stimuli be standardised, the International Commission on Illumination (ICI or CIE[6]) opened the way to establishing an internationally agreed system of objective colour measurement; the monochromatic primary wavelengths selected by the Commission were 700·0 nanometres (orange-red), 546·1 nanometres (green) and 435·8 nanometres (blue-violet).

A straightforward experimental method of observing the full range of additive colour mixtures obtained by combining any three primary lights was demonstrated by James Clerk Maxwell in 1855. In his arrangement, three white light sources are fixed one at each corner of a flat, equal-sided triangular board; placed in front of each, and facing the centre of the board, is a filter transmitting light of an additive-colour primary. Inside the triangle all the colours that can be matched by mixing the three lights are not only seen in the appropriate position but their location can be plotted by drawing a grid or system of linear co-ordinates on the surface of the board, thereby obtaining a rudimentary chromaticity diagram.

However, no matter which three primaries are chosen, there

[6] *Commission Internationale de l'Éclairage*, by which name the Commission is known in Europe.

will always be colours which cannot be included inside the area of the colour triangle; their location cannot therefore be given wholly in terms of positive-value co-ordinates. These colours include, most importantly, the monochromatic spectral wavelengths (other than the selected primaries) which, in relation to the sides of the triangle, are found to lie on a curve known as the locus of spectral colours or *spectrum locus* (Figure 11.2).

To overcome this problem, and avoid the use of negative co-ordinates, a system of colour coding was proposed by the 1931 conference of the CIE based on a set of three 'imaginary stimuli'

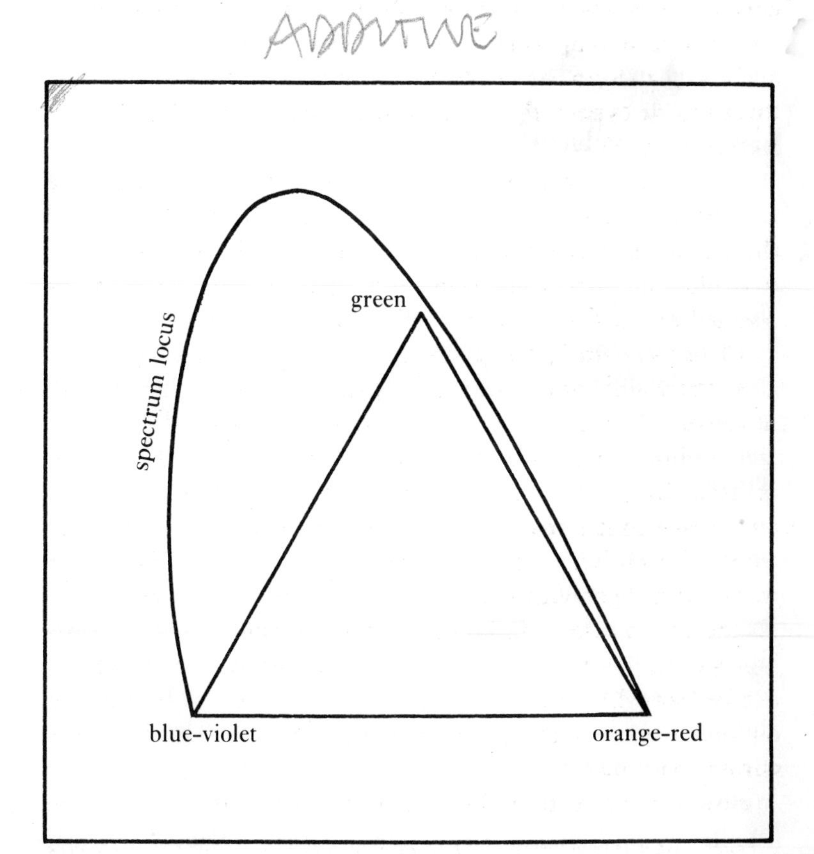

Figure 11.2 Maxwell's colour triangle and the spectrum locus.

possessing in theory far greater purity than Maxwell's real primaries. The CIE imaginary primaries, which are unobtainable physically, are located at the corners of a figure large enough to enclose within it both the Maxwell triangle and the spectrum locus; they are encoded in the CIE system by the letters *X*, *Y* and *Z*, analogous respectively to the orange-red, green and blue-violet primaries of the original colour triangle.

By redrawing the equal-sided triangle as a right-angled triangle it can be transferred to a rectilinear graph. The co-ordinates of such a graph or chromaticity diagram, shown in Figure 11.3, can then be used to fix the position of any point within the area of the graph. The numbers along the curved spectrum locus indicate the wavelengths of the colour spectrum. The straight-line base corresponds to the side of Maxwell's triangular board linking the orange-red and blue-violet primaries, along which are found the purest non-spectral purples; the other two sides of the triangle are included inside the area bounded by the curve. The co-ordinates themselves are denoted by the lower-case letters *x*, *y* and *z*; as the sum of all three is 1 (unity), it is only necessary to quote the first two (from which the third can be deduced if needed).

In an example, the chromaticity of a primary-green filter is located on the CIE diagram by the co-ordinates *x* equals 0·210 and *y* equals 0·710. From this code it is possible to determine both the dominant wavelength and the relative purity of the stimulus. This is done by first finding the chromaticity point of the illuminating light source. For convenience this will be assumed to be the so-called 'equal-energy white' (radiating the same amount of energy throughout the spectrum) located at the point where *x* co-ordinate 0·333 intercepts *y* co-ordinate 0·333. A straight line drawn to connect the white point E with the colour point S is then extended to cross the spectrum locus at the point D in Figure 11.3; this point of intersection fixes the dominant wavelength of the green filter light at 535 nanometres. The distance between the white and the colour points (the line ES on the diagram) divided by that between the white point and the intersection on the locus (the line ED) expresses the relative purity of the stimulus as a

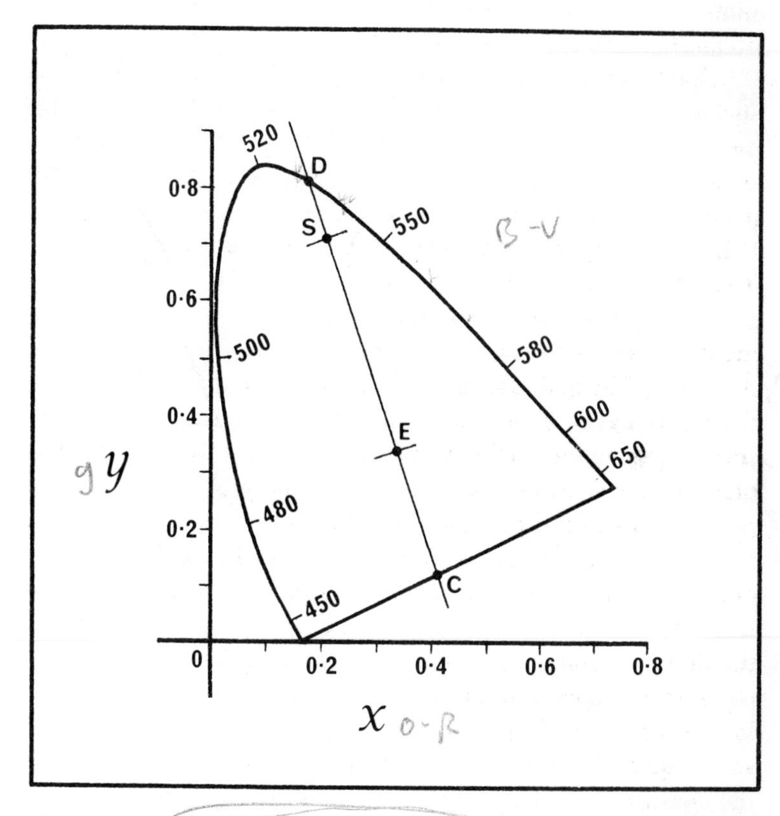

Figure 11.3 The CIE chromaticity diagram. The point D indicates the dominant wavelength of the stimulus S and the ratio ES:ED expresses its relative purity.

fraction, which in this instance is approximately 4/5.

Additionally, by extending the same drawn line on the opposite side of the equal-energy white, it is possible to fix the wavelength *complementary* to the filter light at the point C in Figure 11.3. In the special case of non-spectral purples, it is customary to identify a stimulus lying on or near the straight-line base of the locus by quoting its complementary dominant wavelength as a minus value (though its chromaticity is still expressed by positive x and y co-ordinates).

In practical colorimetry the intensity of a given stimulus can be calculated by adding together the separate intensities of the three primary lights needed to match it. For mathematical convenience the three CIE imaginary primaries (the *tristimulus values*) were defined so that in theory the photometric values of X and Z were zero. The luminance of a given stimulus is consequently indicated entirely by the magnitude of the Y value, which is usually adjusted to express the luminance factor of the stimulus as a percentage ratio.

No one system of specification has yet succeeded in accounting for all the factors which determine a colour response. However, of the many systems which have been proposed, the CIE system has attained international recognition as being probably the most precise and flexible. By quoting the x and y chromaticity co-ordinates plus the Y percentage luminance factor, it establishes a convenient numerical code which is easily interpreted and which provides what is generally accepted for many purposes to be an unambiguous definition of any given stimulus as perceived by an average observer possessing normal colour vision.

CHAPTER 12

The Measurement of Pigments

In the painting, printing and dyeing industries it is often critically important that a colour can be accurately matched and duplicated. Where manufactured articles are composed of materials from several different batches the parts are expected to correspond precisely in colour and the customer will not accept ceramics, fabrics, wallpapers nor articles of clothing in badly matched colours.

Subjective colour measurement is hampered by the fact that visual responses are highly individual. Even when anomalous colour vision does not present a problem, and when no actual sample is available for examination, colloquial colour terms prove to be inadequate; in the absence of defined concepts of apparent colour quality, verbal descriptions are usually imprecise and unreliable.

It is additionally unfortunate that specialists seeking different forms of colour identification have tended to accumulate their own colour terminology. The scientist describes colours with one set of names and the artist with another. Geologists, horticulturists, philatelists and manufacturers of paints, inks, dyes, plastics and cosmetics have each adopted their own colour nomenclature with the result that different terms may identify the same colour or the same term may be used to identify colours which are very different in appearance. Names selected from a commercial housepaint colour chart include *Arcadia*, *Bandbox*, *Charm*, *Dreamboat* and *Enigma*, all apt labels to enchant the amateur decorator but ones which, taken out of context, give no indication of the appearance of the colours they respectively represent.

The need for precision in colour matching has led to the establishment of numerous systems of pigment colour measurement. The first practical attempt was made prior to 1766 by Moses

Harris. Harris was primarily an entomologist and his *Natural System of Colours* (first published in London about 1776), though addressed to artists, arose from a desire to establish a method of classifying the colouring of insects. Another system was devised by ornithologist Robert Ridgway and first published in 1886. Ridgway's arrangement, originally intended to aid the identification of plumage colouration, was in the form of a *colour atlas*, a book of colour samples for reference, in which mixtures derived from 36 principal colours were arranged in an ordered sequence of visually estimated intervals.

When, in 1824, Michel-Eugène Chevreul was appointed *directeur des teintures* at the Gobelins dyeing works in Paris, one of the first tasks he found it necessary to undertake was the rationalisation of the many thousands of dyes then in use at the Gobelins tapestry manufactory. By establishing a system of colour measurement in which mixtures were derived from 12 principal colours, Chevreul succeeded in reducing the number of dyes in use to 1,500 while maintaining rigorous standards of tapestry manufacture.

It was established in Chapter 11 that the physical character of a coloured stimulus can be defined in terms of three variables: luminance, dominant wavelength and purity. These objective dimensions are analogous respectively to *luminosity*, *hue* and *saturation*, three visual attributes which are subjective and not susceptible to physical measurement.

Variation in luminosity, a concept usually reserved for light-emitting sources, is attributable directly to an increase or decrease in the intensity of the source: it indicates the apparent quantity of light emitted. For reflecting surfaces (including pigments), the terms *lightness* or *tone* are usually substituted. While also dependent on the intensity of an illuminating source, lightness is additionally influenced by the light-absorption properties of the material constituting the illuminated surface.

Hue is the attribute of visual perception by which different regions of the spectrum are distinguished and named, thus: red, orange, yellow, green, blue or violet. Hue is not adequate alone to describe fully the subjective quality of a colour since it gives no indication of how much the stimulus appears to deviate from the purity of a spectral light and approach white, black or grey. For this purpose the concept of saturation is introduced.

Saturation is the visual attribute whereby one stimulus is judged to be more or less colourful than another of the same hue. A pigment which appears permeated with colour is said to be saturated or 'fully saturated', while one dulled by admixture of white, black, grey or complementary colour is said to be unsaturated. Two further terms useful to distinguish differences in saturation are *tint*, which properly identifies a mixture of hue plus white (or transparent extender), and *shade*, which identifies a mixture of hue plus black. Surfaces which exhibit no hue response and possess zero saturation, that is, those which vary in lightness only, are called *neutral* and appear colourless when viewed in isolation.

When an observer is presented with a visual stimulus and asked to describe his response, there is commonly a lack of correspondence between what is physically present and what the observer perceives as present. The examination of this phenomenon is embraced in the branch of science known as psychophysics.

It has been found, by psychophysical experiment, that the analogous concepts of luminance and luminosity or lightness, dominant wavelength and hue, and purity and saturation are not as a rule interchangeable. A fundamental difference between physical stimulus and psychological response is exhibited in the spectral sensitivity curves. Those shown in Figure 4.3 (page 27) indicate that, for the same amount of light energy expended, human colour vision is not responsive equally to the different colours of the spectrum. Yellow looks the most luminous colour when the eye is light-adapted and green the most luminous when the eye is dark-adapted. Visual sensitivity diminishes towards either end of the spectrum and eventually disappears beyond about 760 nanometres into the infrared and beyond about 380 nanometres into the ultraviolet.

Another discrepancy between stimulus and response is that a

numerical division of the spectrum into wavelengths does not coincide proportionally with its division into hues. Visual sensitivity to changes of hue is more concentrated in some parts of the spectrum than in others. The effect is most marked in the region of 580 nanometres so that distinct orange-yellow, yellow and yellow-green hues are identifiable in the waveband 550 to 600 nanometres, while wavelengths in the neighbouring band 500 to 550 all appear green. Moreover, there is a range of colours which exhibits perceptible hue but which possesses no corresponding physical wavelength. These colours are the purples (which include pigment-primary magenta) which do not form part of the spectral series but exist as mixtures of red and violet light.

Pigments are intermixed in order to modify one or more of the subjective visual attributes of lightness, hue and saturation. The resulting colour depends on the relative proportions of the pigments which make up the mixture. For this purpose, pigments are generally obtained in the greatest saturation available. The hue of a desired colour is matched by mixing together the two pigments it most nearly resembles; lightness and saturation may then be adjusted by introducing into the mixture suitable amounts of white, black or contrasting colour.

Pigment intermixture is necessary because in practice the painter or printer is unable to keep at his disposal an estimated seven million jars, tubes or cans of colouring material which would be required to match every variation of colour discernible in the visible world. Owing to the remarkable synthesising faculty of the human visual system, by which the unaided eye cannot distinguish monochromatic light from a fusion of lights which yield the same colour appearance, it is possible to match a surprisingly large number of colours from very restricted selections of pigment colours.

In Classical Greek painting, a selection of four pigments, known as the *tetrachrome* palette, was considered sufficient to meet all artistic requirements. It consisted of Red ochre, Yellow ochre, Vine black (a vegetable charcoal) and Chalk white. As Vine black and Chalk white both possess a bluish cast, unsaturated greens and purples were obtainable and it was the claim of Democritus that a total of 819 colours could be derived from this limited selection.

In nineteenth-century France, when artistic colour needs were very different from those of Ancient Greece, Impressionist painters often found it necessary to use relatively large numbers of pigments for mixing, concerned as they were with preserving the full saturation of unmixed paints direct from the tube. An oil palette selected by Paul Cézanne is known to have contained upwards of 15 colours, including Geniune madder, Alizarin crimson, Vermilion, Cadmium yellow, Chrome yellow, Naples yellow, Viridian, Prussian blue, Cobalt blue, French ultramarine, Red ochre, Yellow ochre, Raw sienna, Burnt sienna, Green earth, Peach black (a vegetable charcoal) and Flake white.

Psychophysical experiment has established that a very large number of mixed colours can be derived from three fundamental colours only, these being the subtractive primary colours, magenta-red, cyan-blue and chrome-yellow. Though, in the sixteenth century, Titian declared that a good painter 'needs only three colours', his own palette appears to have contained at least nine, and the earliest verified application of a three-primary pigment system is credited to Jacob Christoph Le Blon who in about 1720 devised a rudimentary method of colour separation for the copying of paintings by copperplate engraving. Le Blon, in his *Coloritto*, first published in London in 1735, proclaimed:

> 'Painting can represent all visible Objects, with three Colours, Yellow, Red, and Blue; for all other Colours can be compos'd of these three, which I call Primitive.'

Though supported by Castel (1740), Harris (c. 1776), Goethe (1810), Runge (1810) and Field (1817), the three-primary pigment theory was not widely accepted by artists in Europe until well into the nineteenth century. Even Delacroix, the most accomplished colourist of his generation in France, was well established in his career before admitting, in 1854, 'at last I have come to convince myself that nothing exists without the three [primary] colours; I used to believe that they were only in certain objects'.

The number of 'Primitive' hues a painter's palette should contain is even now a matter of debate. While it is true that magenta, cyan and yellow can normally be regarded as minimum, as is evident when used transparently in colour printing and colour photography, certain theorists, among them Ewald Hering and Wilhelm Ostwald, have maintained that the spectrum exhibits *four* unitary hues: red, yellow, green and blue. These four colours, known as the *psychological primary colours*, look fundamental and unique in colour quality and psychologically appear to bear no resemblance to one another.

A system of colour measurement based on *five* principal hues was devised by Munsell, who paid particular attention to the mixing of opaque colouring materials. A minimum of *six* hues has been recommended by other authorities,[1] as the range of hues derived from magenta, cyan and yellow colourants currently available is more restricted than the three-primary pigment theory would require.

Isaac Newton, while observing mixtures of spectral lights, noticed that by overlapping the red and violet fringes of the colour spectrum a range of purple lights was obtained which was not present in the spectrum itself. This suggested the idea of linking the two ends of the spectrum and charting the sequence of hues as a chromatic circle or *colour wheel.* Moses Harris applied the concept to pigment colours and plotted in colour and 'in regular, simple, and beautiful arrangement' the diagram Newton had drawn in black and white.

The colour wheel constitutes the basis of almost all systems of pigment colour measurement, including those of Runge (1810), Chevreul (1861), Rood (1891), Munsell (1905) and Ostwald (1916). Runge reduced the colour wheel to three units: magenta opposite green, cyan opposite orange-red, and yellow opposite blue-violet (*see* Colour Plate 3.1). Each couple forms a complementary colour pair the intermixture of which results in colour neutralisation; this

[1] Including the American Martin-Senour *Nu-Hue Custom Color System* devised for industrial use.

is true for all complementary pairs which, for mixtures of pigments, always approach black.

Symbolically, the location of the secondary colours in relation to the primary colours on the colour wheel may be fixed midway along lines connecting the primary points on the circumference of the wheel shown in Figure 12.1. In this arrangement, the *tertiary* colours, each consisting of a mixture of two secondaries, are located midway along lines linking the secondary colour points within the circumference of the wheel.

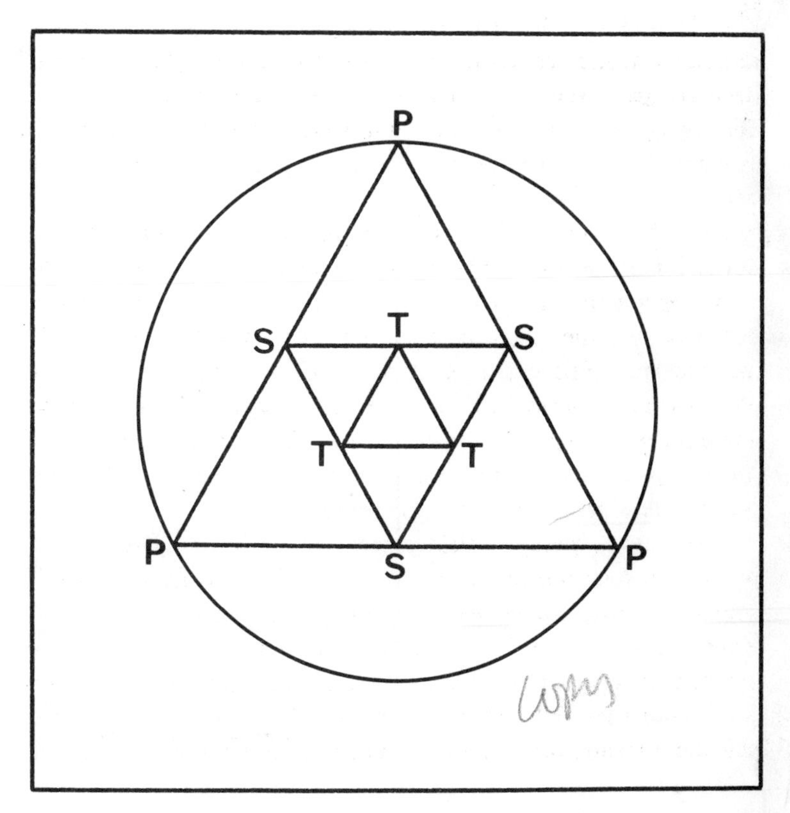

Figure 12.1 Diagrammatic colour wheel, showing primary colours P, secondary colours S and tertiary colours T.

The colour wheel has proved an admirable solution to the problem of representing the hue sequence graphically; but hue is only one of the three distinct ways in which the character of a coloured stimulus can vary. In order to relate hue to lightness and saturation, Runge devised an arrangement in which the colour wheel forms the equatorial ring of a *colour sphere* (Figure 12.2). The vertical axis of the sphere serves to represent the graded lightness scale from white to black; and variations in saturation, derived from each hue, are graded towards the axis, tints upwards to a top, white pole, and shades down to a bottom, black pole.

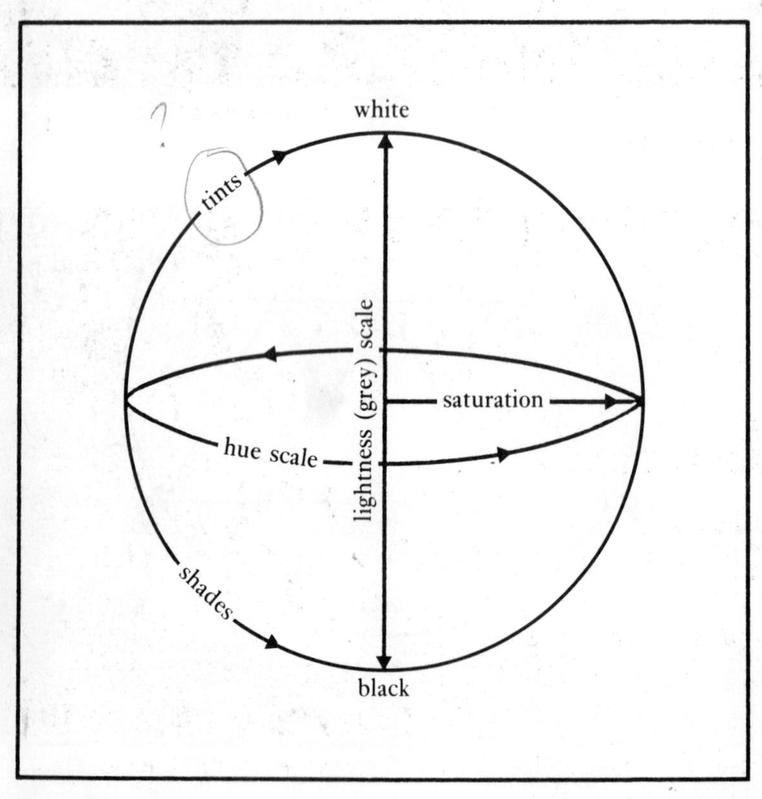

Figure 12.2 Runge's colour sphere.

Colour 5.1 Optical mixing discs. Segmented discs with their resultant mixtures

Colour 5.2 Optical mixing mosaic. Observe the colour blending at different distances

Three-dimensional systems of colour organisation, inspired by Runge's sphere, were subsequently evolved by Ostwald and Munsell, both of whom sought independently to provide for art and industry a simple and easily communicable method of identifying, coding and measuring the appearances of pigment colours regardless of their physical or chemical constitution.

Ostwald's colour wheel consists of a sequence of 24 hues divided into 8 groups of 3, named *yellow*, *orange*, *red*, *purple*, *blue*, *turquoise*, *seagreen* and *leafgreen* (*see* Colour Plate 3.2). In the lightness scale, a standard white sample *a* is linked to a standard black *p* by 13 grey steps, judged visually equal in interval[2] and lettered *b* to *o*; the sequence is usually abridged to 8 steps, *a c e g i l n* and *p*. When organised three-dimensionally, each colour sample is located within a double cone, similar to an arrangement earlier devised by Rood. Hues are distributed around the horizontal circumference of the figure; white is at its top and black at its base.

A vertical cross-section of Ostwald's double cone (Figure 12.3) reveals a pair of precisely complementary leaves, each a single-hue triangle in an arrangement (proposed by Hering) in which hue, white and black are located at its corners. Ostwald adopted as the basis of his system a principle of psychophysics stated by Fechner (1858) that, for all mixtures of pigment colours, *hue content* plus *white content* plus *black content* equals 1 (unity). The standard Ostwald triangle displays 27 gradations of colour, obtained by mixing white and black pigments with an Ostwald hue in standard proportions.

An Ostwald colour specification is given by a number followed by two letters which encode hue content H, white content W and black content B in the sequence HWB. The number denotes hue and two lower-case letters locate the position of the colour sample on a leaf of the double cone. The colour specification *Ostwald 20en* thus indicates a pigment mixture derived from Ostwald hue *20* (mid-seagreen) and consisting of 29·4 per cent

[2] Ostwald here observed a principle stated by Fechner that graded intervals which appear to increase evenly in an *arithmetical* series (1 2 3 4 5 and so on) progress physically in a *geometrical* series (1 2 4 8 16 and so on).

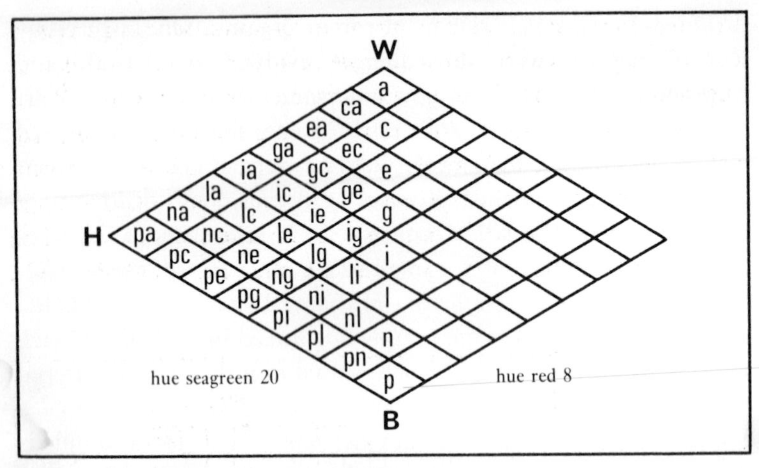

Figure 12.3 Ostwald's colour solid; cross-section. Colours along each line parallel to HB possess the same white content and colours along each line parallel to HW possess the same black content.

hue content, 5·6 per cent white content and 65·0 per cent black content.

An alternative system of colour notation, based on visually uniform spacing, was devised by Albert H. Munsell. Munsell's system seeks to identify pigment colours in terms of the three dimensions hue, *value* and *chroma*; value corresponds to the concept of lightness, and chroma is an estimate of colourfulness corresponding closely to the concept of saturation.

Munsell's colour wheel is composed of 10 segments, consisting of 5 *principal* hues and 5 *intermediate* hues (*see* Colour Plate 3.3). Each segment is divided into 4 quarters, to give 40 actual colour samples. Hue, value and chroma are represented three-dimensionally in the form of a *colour tree* in which 9 actual grey values make up the vertical axis of the tree and link an ideal white *10* at the top of the figure to an ideal black *0* at its base. A cross-section through the colour tree reveals two closely complementary 'branches' each exhibiting a sequence of colour gradations derived from a Munsell hue (Figure 12.4). Unlike Ostwald's arrangement, in which the selection of colour samples is finite, Munsell's system can be augmented by the addition of other colour samples as pigments and dyestuffs of greater chroma are developed.

The letters *R Y G B* and *P* represent the 5 principal hues, red, yellow, green, blue and purple; *YR GY BG PB* and *RP* represent the 5 intermediate hues. A Munsell colour specification is denoted by placing hue H, value V and chroma C in the order H V/C. The code *Munsell 5Y 5/6*, for example, specifies the yellow hue *5Y*,

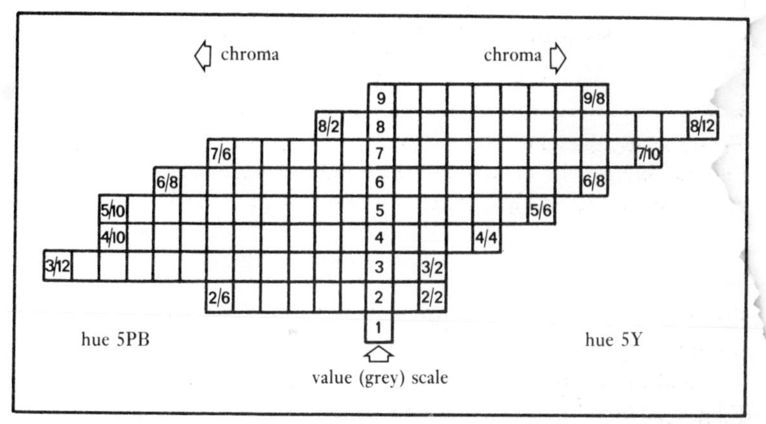

Figure 12.4 Munsell's colour tree; cross-section. All the colours in each horizontal plane possess the same value and all those in each vertical plane possess the same chroma.

middle lightness-value *5*, and chroma 6 steps away from the grey axis in the direction of greatest chroma. A neutral grey, exhibiting neither hue nor chroma, is specified by the value number prefixed by *N*, thus: *N1* to *N9*.

Though no single system of colour measurement can claim world-wide currency, Munsell's decimal notation represents probably the most successful attempt to grade, relate and unambiguously define the appearance of pigment colour samples. The system has been revised and renotated in order to correspond to the CIE chromaticity diagram and, since the 1930s, it has been recognised and approved by the national standards institutions of many countries, including those of the United States, Britain and Japan.

CHAPTER 13

Colour Interaction

'Colours appear what they are not, according to the ground which surrounds them.'

Leonardo da Vinci, *Trattato della pittura*

When two colours are viewed side by side, a change may occur in their appearance in which the contrast between them is heightened or enhanced. This change may be illusory but, for the visual artist, who is dealing with appearances, the change is important. It is useless to devote a great deal of attention to the 'harmonisation' of a selected colour scheme since, as Ruskin pointed out (1857), 'every hue throughout your work is altered by every touch that you add in other places'.

The optical phenomenon in which neighbouring colours are modified in their appearance is known generally as *colour irradiation*. the effect was known to Vasari, who observed, in 1550, 'a sallos colour makes one that is placed beside it more lively; and melancholy and pallid colours make those near them very cheerful and of a certain flaming beauty'.

The first methodical study of colour irradiation was made by Chevreul. It grew out of his discovery that the lack of saturation in certain yarns woven into the fabric of tapestry was due not to imperfect chemical constitution of the dyes used but to the optically irradiative influence of the interwoven colours. Chevreul's original findings were disclosed in a paper read to the French Academy of Sciences in 1828 in which he first stated the principle that, 'when we regard attentively two coloured objects at the same time, neither of them appears of its particular colour, that is to say, such as it would appear if viewed separately, but of a [hue] resulting from its peculiar colour and the complementary colour of the other object'.

Chevreul gave the name *simultaneous colour contrast* to this optical effect, in which adjacent complementary colours will heighten each other's saturation and appear as unlike as possible (*see* Colour Plate 4.1). The contrast effect also occurs in the absence of colour: if, as in Figure 13.1, an identical grey figure is seen against a black and a white ground it will appear simultaneously lighter on the dark ground and darker on the light ground.

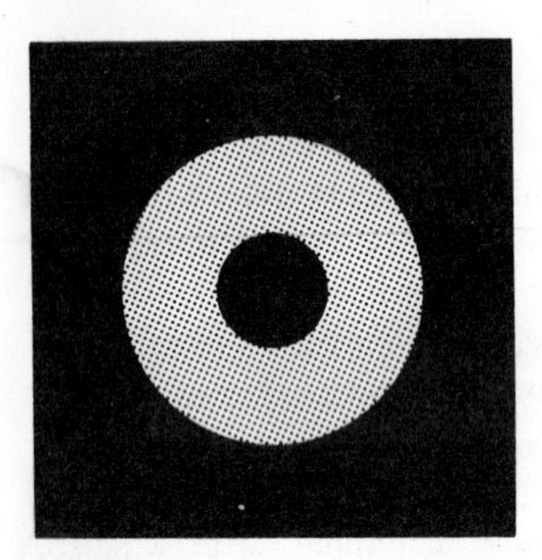
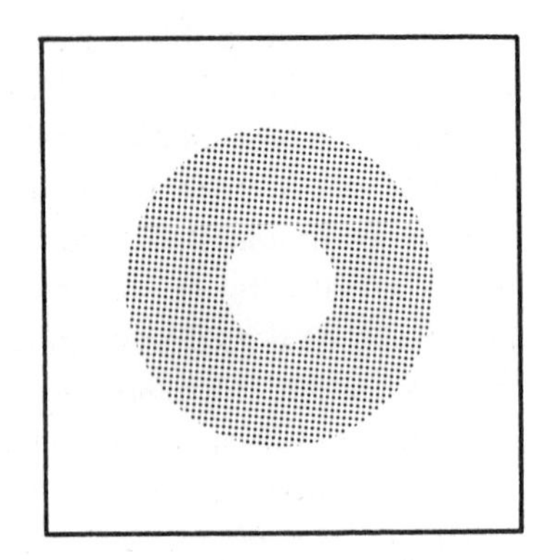

Figure 13.1 Simultaneous lightness contrast. An identical grey figure on black and white grounds.

In expressing his observations in terms of a specific principle, Chevreul devoted considerable attention to the artistic application of the contrast effect. The publication of his researches (1839) was eventually to encourage in a new generation of painters a cultivated awareness of colour contrast and the mutual influencing of juxtaposed colours occurring in the perceived world. For the Impressionist painter, the relationship between a red house and a green tree was to become, primarily, not one between two objects in space but the simultaneous contrast of two dissimilar hues which enliven each other's appearance when placed side by side on the picture surface.

In Paris, following the Impressionist period, painters František Kupka and Robert Delaunay independently evolved methods of pictorial composition the aim of which was to enhance relationships of contrasting colours. Kupka had been exploring contrast phenomena in the context of highly stylised paintings as early as 1907. In an extensive series of pastels based on the theme of *A Woman Picking Flowers* (1909–10: originally numbering over two dozen, five of which are in Paris, National Museum of Modern Art), juxtapositions of unsaturated complementary hues are supported by a framework which divides the picture into broad vertical bands.

Delaunay, in several series of paintings, most importantly the *Windows* (1912), *Discs* (1912) and *Circular Forms* (1912–13), adopted the fragmentation of form characteristic of early Cubism to create 'an architecture of colour' built of 'contrasts set out in such a way as to create structures'. Many of his paintings are non-representational and, lacking a central focus of attention, encourage the eye to wander freely across the picture area, renewing vivid contrasts at each colour boundary. Delaunay, who coined the term *simultanéisme* as a collective name for his work of the period, recalled:

> 'Around 1912–13 I had the idea of a type of painting which would be technically dependent on colour alone, and on colour contrast, but would develop in time *and* offer itself to simultaneous perception all at once.'

The contrast enhancement effect, as Delaunay knew, is also perceived violently when differently coloured stimuli are viewed in succession, in which case the observer perceives a variety of after-image, investigated in turn by Harris, Goethe, Young and Field, and to which Chevreul gave the name *successive colour contrast*.

When an observer fixes his gaze on a coloured stimulus for several seconds, and then directs it away, the sensation of seeing the same colour may persist very briefly as a so-called *positive* after-image. Thereafter, a *negative* after-image is substituted which generally persists for several seconds, its colour corresponding to the complementary colour of the original stimulus; orange-red is replaced by cyan, green by magenta, and blue-violet by yellow (*see* Colour Plate 4.2). As with simultaneous contrast, the effect occurs also in the absence of colour; if the eyes are focused on the

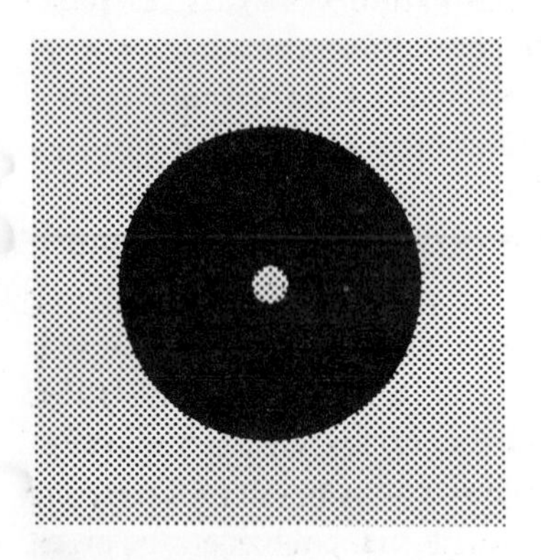

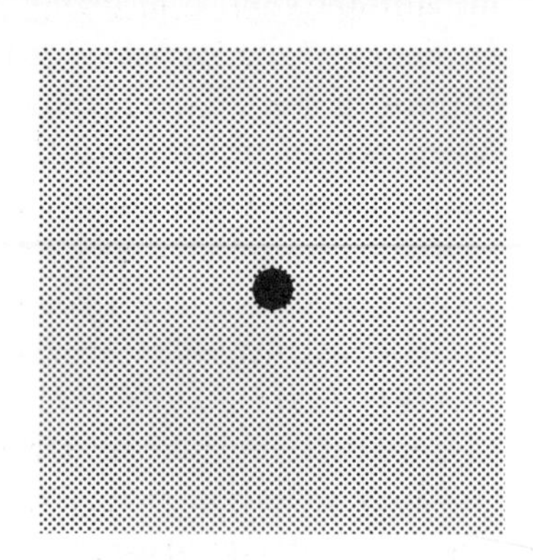

Figure 13.2 Successive lightness contrast.

central point of the black circle in Figure 13.2 for about 10 seconds and then transferred to the central point of the grey square, a whitish 'ghost' circle is perceived at first violently in the centre of the square, and then gradually fades away.

The negative after-image effect, stimulated by prolonged visual fixation, was known to Edvard Munch and exploited in the bizarre colouring of several of his early paintings. In trying to illustrate in an example what seemed 'incomprehensible' to his critics in his use of colour, Munch noted in his diary of 1890:

> 'A billiard table—go into a billiard hall. When you have concentrated intensely on the green cover for a while—look up. How wonderfully reddish everything around will seem. The gentlemen dressed in black now wear crimson-red suits—and the hall has reddish walls and ceiling. After a while the suits become black again. If you want to paint such an impression, with a billiard table, then you must paint them in crimson-red.'

Contemporary artists who have purposely involved successive contrast in the viewing of their work include Larry Poons, Norman McLaren and Robert Breer. In a series of untitled paintings (1962–69) by Poons, coloured eliptical discs are distributed across large canvas panels, of the order of 3 or 4 metres wide, each painted flatly in a single, saturated colour. The eye of the viewer, denied the edge of so large a panel as reference of stability, is stimulated to shift constantly across the picture surface, each saccadic movement inducing complementary after-images which make it difficult, after several seconds viewing, to distinguish the painted discs from those which are wholly illusory.

Hallucinatory effects are further enhanced when duration of viewing time is strictly limited, McLaren and Breer both having made films which rely heavily on the deliberate induction of successive contrast effects. In McLaren's *Blinkity Blank* (1954; 15 minutes), images are scratched into the emulsion layer of otherwise opaque 35-millimetre film and coloured by hand. By scattering 'an image here and an image there', and leaving the greater part of the film blank, McLaren succeeded in creating a remarkably explosive form of animation in which pictures seen for just one forty-eighth of a second persist as after-image illusions.

Of his five-minute film *66* (1966), Robert Breer has explained, 'I hold one image for several seconds and then follow it with an image that is composed around the expected after-image'. In subsequent films, including *70* (1970; 5 minutes), he induces various forms of 'figure-ground reversal' by following a single frame or two- or five-frame grouping of, for instance, a green figure on a red ground with a red figure on a green ground.

In certain circumstances it has been the deliberate intention of the visual artist to curtail or inhibit the expansive irradiation of adjacent colours and instead to promote or enhance the individual quality of each colour in relative isolation. Chevreul, in admiring the 'coloured glass in large Gothic churches' and 'the beauty of the colours of the paintings in flat tints which come from China', had observed how outlines contributed 'to render the impression of the colours stronger and more agreeable'. Rood reserved his admiration for the Alhambra, the palace of the Moorish kings at Grenada, noting how 'colours that differ considerably are prevented by contours from melting into each other and thus giving rise to mixture tints'.

In the conventional type of stained glass window the black strips of leading between each glass fragment, in addition to locking the window panes together physically, ensure minimum optical spreading of the enclosed colours. In a style of painting dubbed Cloisonism, from its resemblance to 'partitioned' enamelwork, the Nabis (1888–99), a group of French painters influenced by Gauguin and nineteenth-century Japanese masters, composed pictures in which broad areas of saturated colour were bounded by blue or black outlines. Some 10 Nabi paintings, corresponding closely to Chevreul's description of 'painting in flat tints', were selected by Tiffany (1895) for transcription into stained glass, a medium to which, because of their characteristic dark contours, they were ideally suited.

In 1946, Fernand Léger recalled a desire, early in his career, to reject the Impressionist practice of placing brushstrokes of vivid colours edge to edge; it had been his wish instead 'to isolate the colour, to produce a very red red and a very blue blue':

> 'It was about 1910 that Delaunay and I began to liberate pure colour in space. Delaunay developed in his own individual way, keeping the relationships of pure complementary colours. I was seeking my own way in the opposite direction—avoiding as much as possible complementary relationships, and developing the force of pure local colours.'

Following an examination of the problem (between 1912 and 1919), Léger found his solution in the employment of the bold black contour. The device is highly characteristic of much of Léger's subsequent output in his later years it is not surprising to find him extending his painting style into the medium of stained glass. Among his designs for the Church of Sacré-Coeur at Audincourt near Belfort in Doubs (1951–52) stained glass slabs were bonded physically within a heavy framework of reinforced concrete.

The black contour also characterises the artwork of Georges Rouault, who was apprenticed to a stained glass painter for five years before starting to paint in watercolour in 1890. Rouault retained a heavily contoured style after adopting the medium of oil paint in 1908 and subsequently returned to his original medium, his designs in glass including windows for the Church of Nôtre-Dame de Toute Grâce at Plateau d'Assy, Haute-Savoie (1945).

Other than using lead bands between panes of differently coloured glass, it was common practice among Gothic glaziers to insert colourless glass beading between adjacent colours which were highly irradiative, as between a red and a blue. In painting, the white contour or *anticerne* is most easily obtained by leaving a narrow strip of unpainted canvas between neighbouring painted areas. As the palettes of many artists became more colourful it became convenient to use a white contour in order to help preserve the individual quality of each colour. The anticerne appears in paintings by Delacroix (after 1855) and also in late paintings by Cézanne. The device was popularised by members of the Fauve group in Paris (1905–8); and Henri Matisse, the principal painter of the group, used both black and white outlines habitually throughout his long artistic career.

In American art of the 1960s the white contour obtained by leaving a strip of unpainted canvas between single-colour areas became a unifying feature in the work of a number of painters, among them Paul Feeley, Morris Louis, Kenneth Noland and Frank Stella; all were working with simple, often symmetrical formats on a large scale. Stella employed the device consistently through the decade and explained, in 1964, 'the color shapes are separated from each other so that no accidental optical mixture can disturb the experience of each shape as such'.

A most exhaustive modern study of optical irradiation was undertaken by Josef Albers both in his own artwork and in the preparation of his *Interaction of Color* (1963), a large folio of prints which drew on his work with students at Yale University (1950–60). The book reflects Albers' own approach to colour teaching which consisted primarily of encouraging the student to develop 'an eye for color' coupled with a fundamental awareness, gained by personal experience and trial and error, that 'in order to use color effectively it is necessary to recognize that color deceives continually'.

CHAPTER 14

Optical Colour Mixing

> 'There is a *mixture of colours* whenever various colours are so divided and combined that the eye cannot distinguish these from each other, in which case the eye receives a single impression.'
>
> *De la loi du contraste simultané des couleurs*, 1839
> Michel-Eugène Chevreul,

If an electric lamp is switched on and off, at first slowly and then more quickly, it will be perceived as a flickering light until a frequency is reached of about 40 flashes per second. At this frequency, the *critical fusion frequency* for the average observer, the flicker appears to cease and is replaced by an impression of a uniformly grey stimulus.

Electric discharge lamps are operated on an alternating current of 50 cycles per second (60 per second in the United States). Such a lamp *appears* to emit light continuously: in fact, its light goes out twice each cycle (100 or 120 times each second) but the intervening dark periods are not appreciated visually. Similar optical illusion occurs in the cinema. Sound film is exposed at 24 frames per second but each frame is projected twice, giving an overall frequency of 48 individual flashes of light per second. This is adequate to fool the observer, who perceives continuous activity on the cinema screen but remains unaware that there are dark intervals between each illuminated picture (that is, that the screen is physically in darkness for half of the time).

The human visual system is incapable of signalling rapid fluctuations of light and dark from the eye to the brain. The visual response to a flash of light tends to persist for up to a quarter of a second after the light has been switched off. This optical phenomenon, which involves the induction of a positive after-image, is known as the *persistence of vision*.

Tony Conrad's 45-minute film, *The Flicker* (1966), consists of black and white alternations which differ only in frequency. From 1 dark pulse per second, frequency increases to 2 and then 4 per second, which appears to be the lower limit of appreciation of 'flicker'. Frequency continues to rise until alternate frames are black and white, before the flicker eventually slows to its original pulse. The resulting stroboscopic effects cause visual-neural disturbances which can induce the perception of vivid coloured patterns (first studied by Fechner) and, in rare cases, photogenic epilepsy.

When, in similar circumstances, an observer is presented with a rapid alternation of two differently coloured stimuli, one colour does not have time to fade before it is replaced by the appearance of the second colour. When the two colours are alternated at a rate above the critical fusion frequency, the stimulus is perceived as continuous: an impression of a single colour is seen which appears to correspond to an additive mixture of the two component colours. The resulting impression depends on two factors: the colour qualities of the component lights and the relative proportions in which they are mixed.

Visual or optical colour mixing occurs when differently coloured lights or film frames are projected in rapid succession either on to a reflective screen or directly into the viewer's eye, or when a mosaic of points or filaments of light are presented which are too fine for the eye to resolve individually at a suitable viewing distance. The mosaic method of optical mixing is employed in systems of colour television in which the picture is composed of tiny luminescent dots or lines emitting orange-red, green and blue-violet lights.

Optical colour mixing also occurs when juxtaposed *pigments* are illuminated by a single or composite source of light. One practical method requires a rotating disc with differently coloured sectors, and another, a static mosaic of differently coloured dots or lines.

In his lecture *On physical optics* (1801), delivered at the Royal Institution, Thomas Young observed:

'The sensations of various kinds of light may be combined ... by painting the surface of a circle with different colours, in a way that may be desired, and causing it to revolve with such rapidity, that the whole may assume the appearance of a single tint, or of a combination of tints, resulting from the mixture of the colours.'

Chevreul, Maxwell and Ostwald, and painters Goethe, Rood, Munsell, Kupka and Albers also found the optical mixing disc an invaluable aid in their respective investigations into colour mixing. The device, customarily known as *Maxwell's disc* (though Maxwell's initial work was performed with a spinning top), is readily assembled by fitting a disc of firm card on to the spindle of a rotary motor or power drill. A radial slit from rim to centre permits the interleaving of several discs to constitute any desired combination of colours through 360 degrees.

An optical mixing disc composed of black and white sectors only, when rotated rapidly, exhibits a uniformly grey appearance the lightness-value of which depends on the proportion of black-sector area to white-sector area.[1] When the sectors are differently *coloured*, the prediction of the resultant colour is more difficult, as a particular resultant colour may be imitated by very different combinations of component colours. Young noted that a mixture composed either of all three additive primaries or a pair of complementaries 'exhibit, when whirled swiftly around, a whitish light'.

A disc divided into orange-red and blue-violet segments, when rotated at high speed, gives an overall impression of magenta-red; green and blue-violet blend to imitate cyan-blue; and orange-red and green blend to imitate yellow (*see* Colour Plate 5.1). Optical mixing by disc would therefore appear to obey the principle of additive colour mixture. However, only one source of light is involved—that of the single or composite illuminating source. Two or more lights are not added together physically so that, while the phenomenon is not subtractive, neither is it truly additive. In consequence, the lightness-value of the spinning disc merely

[1] Owing to a psychophysical principle, outlined by Fechner, a grey which appears for example midway between black and white will be obtained by a ratio of black- to white-area of about 1 : 3 (not 1 : 1).

averages the lightness of its pigment-colour components. The yellow resulting from the optical fusion of red- and green-painted sectors looks dull and unsaturated; and the optical mixture of red-, green- and blue-painted sectors will appear not white but greyish. The same principle underlies optical mixing by the painted mosaic method, to be described.

In his initial work at the Gobelins tapestry manufactory (after 1824), Chevreul found that mutual irradiation of coloured threads could not be ignored particularly when pairs of highly contrasting colours were woven together. The fabric of tapestry consists of interwoven warp and weft threads, the warp extending vertically on the loom and the weft acting as the filling yarn. On the basis of his own experience, Chevreul warned the weaver 'never to admit complementary colours into mixtures which are intended to compose brilliant colours', since, when viewed from a distance, interwoven complementary threads appear uniformly grey, tinged with whichever complementary dominates the mixture (*see* Colour Plate 5.2).

By weaving together threads to make a mosaic too fine for the eye to resolve as individual points of colour, Chevreul found that numerous colour mixtures could be obtained optically from a relatively small number of coloured yarns. With this in mind, he recommended that the mosaic method of optical mixing be adopted commercially in media other than tapestry, including carpet-making, wallpaper staining, glass mosaic and painting.

A technique which John Ruskin considered 'the most important of all processes in good modern oil and water-colour painting' consisted of 'touches or crumbling dashes of rather dry colour, with other colours afterwards put cunningly into the interstices'. Through Ruskin's influence a method of 'using atoms of colour in juxtaposition' had been adopted by members of the Pre-Raphaelite Brotherhood (1849–56) in which small points of pigment were turned individually into a wet ground of Flake white paint and copal varnish.

An oil-painting method of 'colour hatching' practised by Delacroix throughout his career was apparently derived from his early pastel technique. It can be seen in the foreground of *The Massacre at Chios* (1824: Paris, Louvre Museum) in which the flesh colouring of several figures is enlivened optically with small touches of pink, orange, yellow and blue pigment. A similar but bolder method is evident in his murals for the Chapel of the Holy Angels (1854–61) in the Church of S. Sulpice, Paris, completed near the end of his life.

In an analysis of Delacroix's use of colour (1867), Charles Blanc suggested, as an extension of colour hatching, a technique in which small stars or points of paint might contribute to build up large portions of the picture area. This proposal, together with one described by Rood (1879) in which 'different colours are placed side by side in lines or dots', was to inspire the Pointillism adopted by several Impressionist painters in which the artist relies on the visual facility of the viewer to reconstitute such pigment mosaics as optical colour mixtures.

By the mid-1880s, French Impressionism had popularised an approach to pictorial representation the essence of which was 'not to paint the object, but to express sensations' (Cézanne). In paintings of the period by Camille Pissarro, colour was applied to the canvas initially in broad areas over which was introduced a *broken colour* effect in which small brushstrokes of paint were used to imitate the appearance of natural sunlight and shadow.

In Divisionism, a rationalised form of the Pointillist method, hues distributed evenly throughout the colour wheel are isolated on the painter's palette and transferred to the painting surface individually to make a mosaic of small dashes of colour. As each pigment remains unmixed on the picture surface, a much smaller portion of the illuminating light is absorbed than would be the case if the pigments were mixed physically on the palette. The colours constituting the mosaic mix optically and yield quasi-additive colour mixtures which are both lighter and more saturated than those which would be obtained if the same colours were matched by the painter's traditional mode of pigment intermixture.

In a letter to his friend Félix Fénéon, Georges Seurat disclosed that he had been seeking a formula for 'optical painting' (a term he

Colour 6.1
Leonardo da Vinci, *Mona Lisa*, Paris: Louvre Museum

Colour 6.2
Pierre Bonnard, *The Bath*, London: Tate Gallery

preferred to the less-specific label, Neo-Impressionism, coined by Fénéon in 1886) for a period of some six years before commencing *A Sunday Afternoon on the Island of La Grande-Jatte* (1884–86: Chicago, Art Institute). In the painting, which was executed strictly according to the Divisionist principle of optical colour mixing, there is no evidence of the physical intermixture of pigments neither on palette nor canvas. By the consistent application of regularly small dots of colour across the entire picture surface, Seurat endeavoured to emulate, in oil paint on canvas, a method of colour mixing similar to that proposed by Chevreul using woollen threads interwoven on a loom.

The Divisionist palette selected by Seurat consisted of eleven hues, approximately visually equal in interval on the colour wheel, with white available for tinting. A smaller number might have been chosen, but Seurat had found that if the hues are separated by too great an interval then it would not be possible to imitate the nuances of colour thought essential by Rood in any attempt to reproduce colour as manifested in nature. Rood (endorsing an earlier statement by Ruskin) had urged the painter, in his study of natural phenomena, to observe closely the way in which 'the colours pass into each other by gentle and sensible gradations, so that the observer is quite at a loss to say where one ends and another begins'.

Following Seurat's lead, Pissarro had embraced Divisionism enthusiastically in 1884; he was to abandon it only four years later 'after many attempts', finding the method altogether too deliberate and mechanical and 'unable to give an individual character to my drawing'. Among the younger artists, Matisse adopted Divisionism briefly (1904–5) but found it equally restrictive and unsympathetic to 'the rhythm and curve of the line'.

For the greater part of this century (excepting the technical development of colour television and, indirectly, half-tone colour printing), optical mixing by mosaic has been largely ignored by artist and manufacturer alike. Optical painting was eventually revived in the 1960s, following a new climate of interest in the perceptual experience of colour. Its two principal exponents in that decade were Carlos Cruz-Diez and Bridget Riley, both of whom favoured linear motifs. Influenced initially by Land's investigations into colour perception, Cruz-Diez has constructed over 500 *Physichromies* in Perspex and wood, the earlier works (1959–62) employing combinations of red and green only.

Bridget Riley's large-scale use of colour dates from 1966 when colour grading largely replaced the grading of tone which had characterised much of her painting during the preceding five years. In *Late Morning* (1967: London, Tate Gallery), red, blue/green and white vertical bands are distributed across a large rectangular panel. The uniform red is paired with a 10-step blue-to-green gradation and optical blending occurs: the red with the blue induces a magenta glow, and with the green, a dull yellow. The white bands also appear modified: those enclosed by a pair of red stripes are tinged with complementary green, and those enclosed by blue or green stripes with complementary orange.

CHAPTER 15

Colour and Form

As a general rule, active colour interaction in any pictorial scheme will tend to undermine stability of pictorial form, whereas an emphasis placed on form will tend to promote pictorial stability. For Ogden Rood, writing in the late nineteenth century, the distinction between these opposing tendencies was one between 'decoration' and 'painting'.[1] 'In decorative art,' he wrote in *Modern Chromatics* (1879), 'the element of colour is more important than that of form,' whereas 'just the reverse is true in painting: here, colour is subordinate to form.'

Rood's book appeared in the year of the fourth and, in terms of public acclaim and group-identity, the most successful exhibition of French Impressionism, but Rood himself remained staunchly conservative as an amateur painter. At the end of his life (following the turn of the century) he still upheld that the values of academic painting were essentially those of all painting. Yet according to Monet, *Modern Chromatics* (in its French edition of 1881) 'had been of invaluable assistance' to the group of painters who had contributed so much in substituting for the values of academic painting those of 'decoration' derived largely from Oriental sources.

The tendency of everyday visual perception is to bisect the visible world into two parts: *figure* and *ground*. In academic representational art, the painter seeks to render this relationship by employing three main methods of optical illusion. They are chiaroscuro, linear perspective and aerial perspective.

Chiaroscuro, or highlight-and-shadow representation, is a pictorial practice in which the painter imitates the apparent solidity of objects in space on wall, board or canvas which is itself physically

[1] Chevreul had earlier made a similar distinction between the systems of 'painting in flat tints' and 'painting in chiar'oscuro'.

flat. Chiaroscuro illusion is achieved by observing and recording the manner in which whole or partial shadow (umbra or penumbra) play across the surface of objects and define their solid appearance.

'Perspective is the rational law by which experience confirms that all objects transmit their image to the eye in a pyramid of lines' (Leonardo da Vinci). The fundamental assumption of linear perspective is that, although receding parallel lines (such as railway tracks) never meet, there is a point on the horizon at which they appear to do so. In order to co-ordinate the illusion of height, width *and* depth on a flat panel, one, two or more *vanishing points* are located on a horizontal line (corresponding to the eye-level of the viewer) drawn across the picture surface and towards which all receding lines converge.

The term aerial perspective refers to the atmospheric effects responsible for a change in appearance of a sequence of objects which increase in distance from an observer. As they penetrate the atmosphere, the rays of sunlight are scattered by particles of dust in the air. The shorter (blue) wavelengths, affected more than the longer, account for increased blueness in the appearance of shadows cast by distant objects as light travels greater distances through the atmosphere. The scattered blue rays overpower the negligible amount of sunlight reflected from the shadows of distant hills so that they themselves appear sky-blue.

All three methods of representation were used by Leonardo in his celebrated portrait of Lisa Gherardini, called the *Mona Lisa* (1499–1505: Paris, Louvre Museum). The painting, which Vasari justly described as 'an extraordinary example of how art can imitate nature', is uncompromising in its subordination of colour to pictorial form (*see* Colour Plate 6.1).

The picture renders a striking separation between figure and ground, the central portrait figure being defined unambiguously in front of its background. In this arrangement the flesh colours of the figure and the yellow and green of her robe are isolated optically and fail to interact with the blues and greens of the distant hills. Chiaroscuro is rendered by the practice, characteristic of Leonardo's painting method, in which a *lay in* of near-neutral values first establishes the entire lightness scheme of the picture, the artist confining his attention initially to the problems of design and chiaroscuro. Light and shadow, 'blended without lines or borders, in the manner of smoke', imitate the apparent relief and roundness of the figure prior to the application of transparent glazes of local colour over the completed *lay in*.

Despite the separation of figure and ground, both submit to the same organisation of pictorial space. The bases of the colonettes flanking the figure indicate a single, central vanishing point on the horizon, behind the sitter's head. The scaling down of distant objects—a bridge to the right, a winding river to the left—contribute further to the illusion of depth and perspective recession. In the background, which exhibits a fine example of aerial perspective, reddish earth colours occupy the near-distance and the vista recedes to shrouded blue hills and sky.

In Northern European painting of the twentieth century the trend has been increasingly one of the rejection of pictorial depth-illusion and the substitution of a feeling for the decorative values of an art in which 'the main object is to beautify a surface by the use of colour rather than to give a representation of the facts of nature' (Rood). In an example of this trend, Pierre Bonnard's portrait of Mme. Bonnard, entitled *The Bath* (c. 1925: London, Tate Gallery), the flatness of the picture surface is affirmed, and depth-illusion contradicted, as a result of the subservience of form to colour (*see* Colour Plate 6.2).

The paint is applied to the canvas *alla prima* as a broad patchwork of interlocking colour areas. Shape, contour and panel-size are defined as the picture is painted, thereby dispensing with 'the academic rule' by which a painter is required to 'draw first, then colour it in' (Nolde). The consequence of this approach is a pictorial ambiguity which permits the painting to be interpreted either as a view in depth or as a quilt-like pattern of colours described by Bonnard as 'a sequence of spots which join together and end up forming the object'.

The Bath exhibits no extremes of light and dark, which

might otherwise serve as stable points of focus; instead a limited tonal range, typical of much of Bonnard's work, encourages the eye to explore the picture as a whole and move with ease between 'figure' and 'ground'. The blurring of the edges of the painted colour areas also helps promote active interaction of colour across the picture surface.

Both bath and reclining figure are cut off abruptly on either side of the picture, omitting a further clue by which the spatial dimensions of the scene might have been defined. No perspective framework suggests Leonardo's 'pyramid of lines': the major divisions of the picture are all horizontal and their colouring largely contradicts the convention of aerial perspective (in which reds signify nearness and blues distance). Finally, the absence of detail and texture, owing partly to the submergence of the bather, removes other key clues customarily used in academic representation.

The paintings by Leonardo and Bonnard might be taken to embody two opposing trends within the realm of visual art. They symbolise, on the one hand, a highly intellectual art of colour subordinated to form, and on the other, a more intuitive art of form subordinated to colour. Between the two extremes lies the potential for systematic exploration of colour and form at a creative, experimental level.

In 1961, Johannes Itten recalled that 'as early as 1917 I made the students use the chessboard division for most exercises in order to free the study of colour effects from the associations of form'. Itten had found that the chessboard design or grid (Figure 15.1) had the optical effect of weakening the element of pictorial form and securing a correspondingly strong interaction of colour across the area of the picture.

By 1920 Itten was teaching at the Weimar Bauhaus and Josef Albers (though born in the same year as Itten) was one of his students. Albers' own systematic use of colour dates from about 1925 and by the early 1930s he was producing visual works in which form remained constant or uniform while colour was variable. Albers referred to the concept as 'the art of given form' and

Figure 15.1 A chessboard design.

in his own career proved himself to be one of its most outstanding exponents. Between 1949 and his death in 1976 he produced a vast series of paintings and prints collectively entitled *Homage to the Square*, based on an arrangement of circumjacent squares (Figure 15.2). Other than the 'target' designs of Noland, Johns, Sedgley and others, it is difficult to imagine a pictorial format which permits so great an emphasis to be placed on colour interaction, an effect heightened when examples of Albers' *Homage* series are viewed side by side.

If the concept of 'given form' were extended to encompass

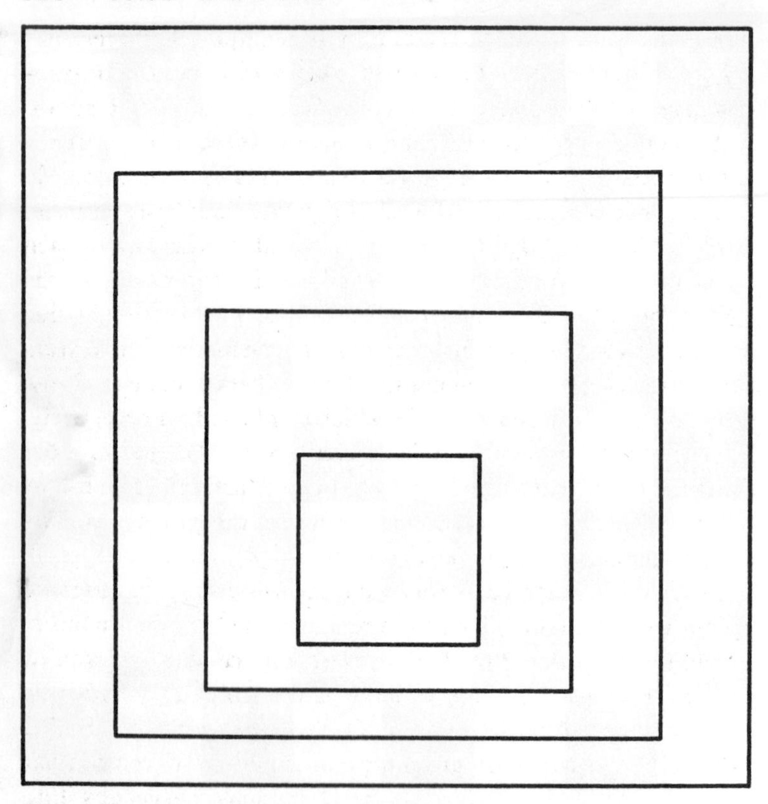

Figure 15.2 *Homage to the Square* (Albers).

representational art, its origin might be credited to Monet who in the 1890s painted several series of pictures intending to record the transitory colour appearance of a particular subject at different times of the day. The two most notable sets of paintings are those of haystacks (1890–91; 15 of which were exhibited together in 1891) and Rouen Cathedral (1892–95; 20 of which were exhibited together in 1895). Evaluated as a series, rather than as individual works in isolation, colour assumes an exaggerated importance while form is correspondingly subordinated. In 1969, images from both series were selected by Roy Lichtenstein for further serialisation in the medium of lithography. Lichtenstein's *Haystacks* were editioned in a series of seven different colour combinations and his *Cathedral* in a series of six colour combinations.

A trend counter to 'given form' might be termed an art of 'given colour'. Among the modern exponents of such an art, in which colour is standardised to permit greater emphasis on form, was Piet Mondrian. By 1917 Mondrian was advocating an approach to pictorial composition which 'must find its expression in the abstraction of all form and colour, that is, in the straight line and the clearly defined primary colour'. In the following two years Mondrian occupied himself both with a number of coloured chessboard grids and with linear motifs which eliminated colour entirely. Throughout the remainder of his career (until 1944), and in order to release formal considerations from the distraction of free colour selection, he chose to restrict his palette to 'the primary colours (red, yellow and blue) and the non-colours (white, black and grey)'.

In 1963, Ellsworth Kelly selected three colours, similar to those chosen by Mondrian, which have remained uniform in a number of paintings entitled *Red Yellow Blue*. The colours are applied flatly to three-part divisions of large panels which vary widely in size and shape. Alternatively, in his *Spectrum* paintings (*II–VI*, 1966–69), consisting of groupings of equal-sized rectangular panels, it is the formal framework which remains consistent within each painting, while colour sequence varies.

One might similarly discern opposing trends in the work of Frank Stella. In the earliest phase of his career, Stella produced a number of groups of single-colour panels of painted stripes; the first (1958–60) employed black housepaint, the second (1960) aluminium paint, and the third (1960–61) copper paint. The colour remained uniform in each group while Stella worked methodically —but at no time predictably—through variations based on a number of simple geometrical forms.

In 1967 Stella began his 'Protractor' series, initially conceived as a set of 93 paintings based on three sets of predetermined configurations, in which form remained relatively uniform throughout each set while colour was varied. Of the series, the so-called 'Interlace' paintings consist of semicircular motifs worked out with compass and straightedge. Their visual impact effectively destroys stability of pictorial form by leading the eye of the viewer back and forth along seemingly interwoven arcs of colour. Robert Delaunay has been acknowledged by Stella as a major influence in the Interlace paintings and one might recognise similarity of artistic intent, most obviously in Delaunay's *Circular Forms* (1912–13) and *Rhythms* (1930–38).

Delaunay has used colour to embellish form while simultaneously using simplicity of form to enhance colour interaction.

Figure 15.3 *Grafias* (Feeley).

Feeling himself 'driven by a need for movement' he sought a new way of fulfilling the prime requirement of decorative art: that of suppressing the dimension of pictorial depth and, by prolonging the element of viewing time, animating the viewer's eye sufficiently to explore the entire picture area. By adapting Cézanne's fragmentation of form into stylised chessboard grids he found he could indeed lure the eye across the canvas in an extended but vain search for a stable point of focus on which it might rest.

Paul Feeley, in trying to preserve the segregation of figure and ground while also affirming the decorative potential of the flat picture panel, composed a number of arrangements of quatrefoil shapes (1954–66). In his *Grafias* (1965: New York, Sheldon Solow Collection), for example, either the light or the dark (the yellow or the blue) shapes become either figure or ground but neither figure and figure nor ground and ground at the same time (Figure 15.3). A visual ambiguity is generated which denies stability of form by successively supporting and undermining figure-and-ground perception. The result is a lively interaction of all the elements of the painting and the simultaneous involvement of colour *and* form.

All visual art entails a synthesis of colour and form—one each artist must examine individually in the context of his own aims, interests and artistic requirements. Serially, this synthesis is open to exploration. It can be expressed in terms of two complementary tendencies: a series of works in which colour is uniform while form is variable will throw emphasis on form; while one in which form is constant while colour is variable will throw emphasis on colour.

PART IV

Appendices

APPENDIX 1

Glossary

Absorption of light. The retention or subtraction by a material of part of the light falling on to it. The proportion absorbed depends on the specific molecular structure of the surface; and the colour the surface appears is given by the dominant wavelength of the light not absorbed but conveyed to the eye by *reflection* from or *transmission* through it.

Additive mixture of colours. Colour fusion obtained by combining or adding light from two or more differently coloured light sources. This is achieved by directing the coloured lights to enter the eye either simultaneously or in rapid succession, or in the form of a mosaic too fine for the eye to resolve. Compare *Subtractive mixture* of colours.

Amplitude of a wave. The greatest displacement from zero position (equilibrium) or the extent of oscillation experienced by a particle of a medium as a wave is propagated through it; for a moving pendulum this is equal to half the length of its swing. Amplitude corresponds to the *intensity* of the energy carried by a wave.

Artificial. Factitious or manufactured; not occurring in nature. Compare *Natural.*

Chromaticity of a stimulus. Physical colour quality of a stimulus, consisting of dominant *wavelength* and *purity*.

Colour. The subjective interpretation by the human central nervous system of sensations promoted in the eye by the reception of radiation of the waveband 380 to 760 nanometres, approximately, which identifies the visible solar spectrum. See *Hue* and *Wavelength*.

Complementary colour. Complement, true contrasting colour or minus-colour. A colour which, by additive or subtractive mixture with a given colour, blends to yield a *neutral* appearance.

Dyestuff. The chemical colouring agent of dye; it is soluble in its liquid vehicle and becomes part of the material it colours. Compare *Pigment*.

Electrode. The conductor by which an electric current enters or leaves a conducting region.

Filter, optical. Light filter or colour filter. A device consisting of a transparent or translucent material which absorbs light selectively while transmitting light of a specific waveband. As a rule, a filter permits the transmission of light of its apparent colour and absorbs the light of its complementary colour. Permanent dyestuffs are usually used to colour materials such as glass, gelatin or cellulose acetate.

Fluorescence. Luminescence emission by a material during the absorption of incident energy at another, almost always shorter, wavelength. Compare *Phosphorescence*.

Frequency of a wave. The number of oscillations or cycles of a wave motion per second; that is, the number of complete waves passing a fixed point per second. As the velocity of light is constant in each medium, the frequency of a wave is inversely proportional to its wavelength. The SI unit of frequency is 1 hertz (1 cycle per second). The frequencies of light waves, measured in 10^{12} hertz (million-million cycles per second) are as follows:

Red	428 to 463
Orange	463 to 513
Yellow	513 to 522
Green	522 to 610
Blue	610 to 712
Violet	712 to 750

Hue of a stimulus. Apparent colour quality, independent of *saturation*; the subjective appreciation of dominant wavelength. The attribute of colour perception by which is distinguished the different parts of the spectrum of colour, namely, red, orange, yellow, green, blue, violet or purple (which is a mixture of red and violet light).

Illumination of a surface. Illuminance. The amount of light falling on a surface per unit area of that surface per second. The SI unit of illumination is 1 lux (1 lumen per square metre).

Incandescence. The emission of light by a material solely because it is heated. Red or white heat, as emitted by the sun, a flame or a filament lamp. Compare *Luminescence*.

Intensity of a light source. Luminous intensity or candlepower. The amount of light emitted per second by a source in a given direction within a cone of unit solid angle. The SI unit of luminous intensity is 1 candela (1 lumen per steradian).

Light. Radiant energy capable of promoting the sensation of vision. See *Stimulus, visual*.

Lightness of a stimulus. Tone, value or grey scale. Apparent brightness of a surface; the subjective appreciation of luminance. The attribute of visual perception by which one surface may appear to reflect a larger or smaller proportion of its incident light than another.

Luminance of a stimulus. Photometric brightness. The amount of light reflected or transmitted by a source or surface in a given direction per unit area of projection per second. The SI unit of luminance is 1 nit (1 lumen per steradian).

Luminescence. The emission of light by a material without sensible heat. See *Fluorescence* and *Phosphorescence* and compare *Incandescence*.

Luminosity of a stimulus. Apparent brightness of a light source; the subjective appreciation of luminance. The attribute of visual perception by which one stimulus appears to convey more light to the eye than another.

Luminous flux. Light flux. The total amount of light emitted in all directions from a source per second. The SI unit of luminous flux is 1 lumen.

Monochromatic. Of one wavelength only.

Nanometre. Millimicron. The SI unit of length equal to 10^{-9} (one thousand-millionth) of a metre, or 10 Ångström units. The nanometre is now customarily used to measure wavelength of light. Abbreviation *nm*.

Natural. Native or genuine; occurring in nature. Compare *Artificial.*

Neutral. Achromatic or colourless. Of the white-grey-black scale of values, exhibiting no colour response.

Objective. Physical or phenomenal; independent of an observer. Compare *Subjective.*

Opaque. Incapable of transmitting light. Optically not transparent nor translucent.

Organic. Derived from organisms, that is, from an animal or vegetable source. Typically containing carbon as one of its elements.

Phosphorescence. Luminescence emission by a material absorbing energy at another, almost always shorter, wavelength which continues for some time after the input of energy has ceased. Compare *Fluorescence.*

Pigment. The chemical agent which imparts colour to most paints and inks; colouring powder dispersed and suspended in its vehicle in which it is relatively insoluble. Compare *Dyestuff.*

Primary colours. Fundamental or unitary hues or wavelengths from which other colours may be derived by mixture but which cannot themselves be imitated by mixture. The primary colours of light, employed in the *additive mixture* of colours, are orange-red, green and blue-violet; they may be represented respectively by the wavelengths 650, 530 and 460 nanometres. The primary colours of pigments, employed in the *subtractive mixture* of colours are magenta-red, chrome-yellow and cyan-blue, respectively —505, 575 and 480 nanometres. The four perceptual or psychological primaries, which give the appearance of being fundamental are red, yellow, green and blue, which may be represented by the wavelengths 615, 575, 545 and 475 nanometres.

Purity of a stimulus. Excitation purity. Physical colour quality independent of dominant wavelength. That part of the *chromaticity* of a stimulus by which a monochromatic light is distinguished from one of the same dominant wavelength which approaches a neutral or white appearance.

Radiation. Electromagnetic radiation. Radiant energy emitted by a source, such as the sun, believed to consist of a continuous electrical and magnetic wave travelling as a discontinuous series of energy packets. Radio, infrared, light, ultraviolet and X-ray are names given to different wavebands of radiation.

Reflection of light. The returning of light falling on a surface. For a polished surface, such as a mirror, reflection is specular and the angle of incidence of the light ray is equal to its angle of reflection; for a matt surface, such as blotting paper, reflection is diffuse or irregular and light is returned at many different angles, regardless of the angle of incidence of the light. Compare *Transmission.*

Representation in art. Pictorial likeness. The illusion of solid objects in spatial recession, achieved by highlight-and-shadow rendering and/or the use of linear and aerial perspective.

Saturation of a stimulus. Colour intensity, colour weight or freedom from grey. Apparent colour quality or colourfulness of a stimulus, independent of hue; the subjective appreciation of purity by which is distinguished the appearance of a colourful stimulus from one of the same hue which approaches a colourless or neutral appearance.

Secondary colour. A colour obtained by mixing together two primary colours.

Spectrum of colour. Visible radiation of the waveband 380 to 760 nanometres, approximately, occurring when white light is dispersed into its constituent colours. An incandescent solid or liquid emits a 'continuous' or 'band' spectrum exhibiting all the wavelengths within its range; an excited gas or metal vapour emits a discontinuous 'line' spectrum consisting of selected light wavelengths only.

Stimulus, visual. A change in external or internal energy which excites the nervous system sufficiently to arouse a visual response in the observer.

Subjective. Psychological or apparent; as judged personally by any one observer. Compare *Objective*.

Subtractive mixture of colours. Colour blending or fusion which is the result of the absorption or subtraction by a material or materials of incident light from a single or composite source. This is usually accomplished by the superimposition of differently coloured filters or by the physical intermixture of differently coloured paints, inks or dyes. Compare *Additive mixture* of colours.

Translucent. Partially transparent. Permitting the transmission light but with some scattering of light waves, so that an object can be seen through it but not clearly; as, for example, etched or frosted glass. Compare *Transparent*.

Transmission of light. The passage of light through an optically transparent or translucent medium, such as air, glass, water or frosted glass. Compare *Reflection*.

Transparent. Permitting the transmission of light without noticeable scattering of light waves, so that an object can be seen through it clearly; as, for example, clear glass. Additionally, a 'transparency' is a photographic positive which is viewed by transmitted light. Compare *Opaque*.

Vehicle. Medium or tempera. The inert liquid in which pigment is carried insolubly, as in most paints and inks, or in which a dyestuff is soluble.

Velocity of a light wave. Constant speed of light in a specific direction. The distance travelled by a light wave per second. In a vacuum this is 299,792 kilometres per second (186,282 miles per second). In other transparent or translucent media the speed of light is given by its speed in a vacuum divided by the refractive index of the medium. Velocity, *frequency* and *wavelength* are related to each other by the equation: velocity equals frequency times wavelength.

Wavelength of a wave. The distance between corresponding points of two adjacent waves. Wavelength may be used to denote the physical colour quality of a spectral light independent of its purity. Colours other than spectral lights are fixed numerically by quoting their 'dominant wavelength', this being the spectral wavelength to which the stimulus most closely corresponds; a non-spectral purple is fixed by quoting the wavelength of its complementary colour as a minus value. The wavelengths of the spectrum of colour, measured in nanometres, are as follows:

Red	760 to 647
Orange	647 to 585
Yellow	585 to 575
Green	575 to 491
Blue	491 to 424
Violet	424 to 380

APPENDIX 2

Biographies

Maurice Agis (1931). English painter and sculptor; born in London. Working largely in London since 1960.

Josef Albers (1888–1976). American painter, designer, graphic artist and poet; born at Bottrop, Westphalia. Bauhaus master at Weimar, Dessau and Berlin 1923–33, then based at Beria, North Carolina, and New Haven, Connecticut. Publications include *Interaction of Color*, New Haven and London 1963.

Michelangelo Antonioni (1912). Italian writer and film director; born in Ferrara. Based in Rome since 1939.

Bernard Aubertin (1934). French painter and light artist; born at Fontenay-aux-Roses, Hautes-de-Seine. Working largely in Paris since 1956.

John Logie Baird (1886–1949). Scots inventor and electrical engineer; born at Helensburgh, Strathclyde. Based in Glasgow, Trinidad, Hastings, Sussex, and London (after 1924).

Jordan Belson (1926). American painter, film-maker and light artist; born in Chicago. Based in San Francisco since 1947.

Margaret Benyon (1940). English painter and holographer; born in Birmingham. Based in London, Nottingham, Glasgow and (since 1976) Canberra, Australia.

Charles Blanc (1813–82). French writer and editor; born at Castres, Tarn. Based in Paris. Publications include *Grammaire des arts du dessin*, Paris 1867.

Pierre Bonnard (1867–1947). French painter, sculptor and graphic artist; born at Fontenay-aux-Roses, Hautes-de-Seine. Active after 1888 in Paris and then Le Cannet, Alpes-Maritimes.

Sandro Botticelli (1445–1510). Florentine painter. Active largely in Florence after 1465.

Pierre Bouguer (1698–1758). French natural philosopher, astronomer and mathematician; born at Croisic, Brittany. Based in Paris after 1735. Publications include *Traité d'optique sur la gradation de la lumière*, Paris 1729.

Robert Breer (1926). American painter, sculptor and film-maker; born in Detroit, Michigan. Based in Paris 1949–59 and then Palisades, New York.

Fred Burrell (1932). American photographer; born in New York City. Based in New York since 1954.

Louis-Bertrand Castel (1688–1757). French natural philosopher, priest and mathematician; born at Montpellier, Herault. Based in Paris after 1720. Publications include *Nouvelles expériences d'optique et d'acoustique*, Paris 1735 and *L'optique des couleurs*, Paris 1740.

Edgar Cayce (1877–1945). American psychic, healer and photographer; born at Hopkinsville, Kentucky. Based at Selma, Alabama, 1912–23 and then Virginia Beach, Virginia.

Paul Cézanne (1839–1906). French painter and graphic artist; born in Aix-en-Provence. Active largely in Paris 1861–86 and then at Aix.

Marc Chagall (1887). French painter, sculptor, designer and graphic artist; born in Vitebsk, Byleorussia. Active after 1910 in Paris, Vitebsk, New York and (since 1950) Vence, Alpes-Maritimes. There is a Chagall Museum at Nice.

Michel-Eugène Chevreul (1786–1889). French chemist; born at Angers, Maine-et-Loire. Based in Paris after 1803. Publications include *De la loi du contraste simultané des couleurs*, Paris 1839 and *Des couleurs et leurs applications aux arts industriels*, Paris 1864.

Chryssa (Chryssa Mavromichali) (1933). American painter, sculptress and light artist; born in Athens, Greece. Based in New York since 1957.

Tony Conrad (1940). American musician and film-maker; born at Concord, New Hampshire. Working largely in New York since 1962.

Adrian Cornwell-Clyne (Adrian Bernard Klein) (1892–1969). English writer and light artist; born in London. Based in London and Hove, Sussex. Publications include *Colour Music, The Art of Light*, London 1926 (revised 1937 as *Coloured Light, An Art Medium*) and *Colour Cinematography*, London 1936.

Lloyd Cross (1934). American holographer and optical engineer; born at Flint, Michigan. Working largely in San Francisco since 1971.

Carlos Cruz-Diez (1923). Venezuelan painter, designer and light artist; born in Carácas. Working largely in Carácas 1945–60 and then Paris.

Salvador Dali (1904). Spanish painter, designer and film-maker; born at Figueras, Catalonia. Based in Madrid 1921–28 then Paris, New York City and Cadaqués.

Ronald Davis (1937). American painter and sculptor; born at Santa Monica, California. Working in Los Angeles 1965–73 and then Malibu, California.

Edgar Degas (1834–1917). French painter, sculptor and graphic artist; born in Paris. Active in Paris after 1862.

Eugène Delacroix (1798–1863). French painter; born at Charenton-le-Pont, Val-de-Marne. Active largely in Paris after 1821. There are large collections of his work in Paris (Eugène Delacroix Museum and Louvre Museum).

Robert Delaunay (1885–1941). French painter and graphic artist; born in Paris. Active largely in Paris after 1908. There is a large collection of his work in New York City (Solomon R. Guggenheim Museum).

Democritus of Abdera (c. 460–361 B.C.). Thracian natural philospher and mathematician. Active in Teos and Egypt.

Walter Elias Disney (1901–66). American animator and film producer; born in Chicago. Founded Walt Disney Productions at Burbank, California, 1927.

Ivan Dryer (1939). American film-maker and astronomer; born in California. Founded Laser Images Incorporated at Van Nuys, California, 1971.

Louis Ducos du Hauron (1837–1920). French physicist and inventor; born at Langon, Gironde. Based in Paris and Agen, Lot-et-Garonne. Publications include *Les couleurs en photographie*, Paris 1869.

George Dunning (1920–79). Canadian animator and film director; born in Toronto, Ontario. Based in Ottawa, Burbank, California, and London (after 1957).

Thomas Alva Edison (1847–1931). American inventor and electrical engineer; born at Milan, Ohio. Based at Menlo Park, New Jersey, 1876–87 and then West Orange, New Jersey.

Gustav Theodor Fechner (1801–87). German physicist and psychologist; born at Gross-Särchen, Lusatia. Based in Leipzig after 1817. Publications include *Elemente der Psychophysik* (2 volumes), Leipzig 1860.

Paul Feeley (1910–66). American painter and sculptor; born at Des Moines, Iowa. Based in New York City 1931–39 and then Bennington, Vermont.

Félix Fénéon (1861–1944). French writer; born in Turin. Associated with Symbolism in Paris. Publications include *Impressionnistes*, Paris 1886.

George Field (c. 1777–1854). English chemist and inventor; born at Berkhampstead, Hertfordshire. Based in London after 1797. Publications include *Chromatics*, London 1817 and *Chromatography*, London 1835.

Niels Ryberg Finsen (1860–1904). Danish physician; born in the Faroe Islands. Based in Copenhagen after 1890. Nobel Prizewinner 1903.

Dan Flavin (1933). American writer and light artist; born in Jamaica, New York. Working at Cold Springs and Garrison, New York, since 1961.

Dennis Gabor (1900–79). British physicist; born in Hungary. Based in Berlin 1924–33 and then Rugby, Warwickshire, and London. Nobel Prizewinner 1971. Publications include *Inventing the Future*, London 1963.

Antonio Gaudí (1852–1926). Spanish architect; born at Reus, Catalonia. Active largely in Barcelona after 1877.

Paul Gauguin (1848–1903). French painter, sculptor and graphic artist; born in Paris. Associated with Impressionism in Paris 1884–86 and then Synthetism in Brittany; later based in Tahiti and Dominica.

Hugo van der Goes (c. 1440–82). Flemish painter; born in Antwerp. Active in Ghent 1467–77 and then Brussels.

Johann Wolfgang von Goethe (1749–1832). German dramatist, poet and painter; born in Frankfurt-am-Main. Based in Weimar after 1775. Publications on colour include *Beiträge zur Optik*, Weimar 1791–92 and *Zur Farbenlehre*, Tübingen 1810.

Vincent van Gogh (1853–90). Dutch painter and graphic artist; born at Groot-Zundert, Brabant. Active after 1881 in The Hague, Paris and Arles, Bouches-sur-Rhône. There is a large collection of his work in Amsterdam (Vincent van Gogh Foundation).

Jean-Baptiste Guimet (1795–1871). French chemist; born at Voison, Isère. Based in Toulouse until 1834 and then Lyons.

Richard Hamilton (1922). English painter, designer and graphic artist; born in London. Working in London since 1948 and Newcastle-upon-Tyne 1953–66.

Moses Harris (1731–83). English entomologist and engraver. Publications include *The Aurelian, or, Natural History of English Insects*, London 1766 and *Natural System of Colours*, London c. 1776.

Hermann von Helmholtz (1821–94). Prussian psychologist, mathematician and theoretical physicist; born in Potsdam, Brandenburg. Based in Berlin, Königsberg, Heidelberg and Charlottenberg. Publications include *Handbuch der physiologischen Optik*, Leipzig 1856–66 (first complete edition 1867).

Ewald Hering (1834–1918). German physician, psychologist and physiologist; born at Altgersdorf, Saxony. Based in Leipzig, Vienna and Prague. Publications include *Zur Lehre vom Lichtsinne*, Vienna 1878.

Charles Edward Iredell (1877–1961). English consulting surgeon. Based in London. Publications include *Colour and Cancer*, London 1930.

Johannes Itten (1888–1967). Swiss painter, sculptor, writer and graphic artist; born at Südern-Linden, Thun. Bauhaus master at Weimar 1919–23, then based in Berlin, Krefeld and Zürich. Publications include *Kunst der Farben*, Ravensburg 1961.

Paul Jenkins (1923). American painter, sculptor and graphic artist; born in Kansas City, Missouri. Working in New York and Paris since 1948.

Jasper Johns (1930). American painter, sculptor, designer and graphic artist; born in Allendale, South Carolina. Working in New York City and Stony Point, New York, since 1952.

Peter Jones (1939). English painter and sculptor; born in Portsmouth, Hampshire. Based in London since 1958.

Wassily Kandinsky (1866–1944). Russian painter, writer and graphic artist; born in Moscow. Active in Munich and Moscow 1896–1921, Bauhaus master at Weimar, Dessau and Berlin 1923–33, then based in Paris. Publications include *Über das Geistige in der Kunst*, Munich 1912. There are large collections of his work in Munich and in New York City (Solomon R. Guggenheim Museum).

Ellsworth Kelly (1923). American painter, sculptor and graphic artist; born at Newburgh, New York. Based in Paris 1948–54 and then New York City and Chatham, New York.

Johann Kepler (1571–1630). German natural philosopher, astronomer and mathematician; born at Weil-der-Stadt, Württemberg. Based in Graz 1594–98 and then Prague, Linz and Ulm. Publications include *Ad vitellionem paralipomena*, Frankfurt-am-Main 1604.

Ernst Ludwig Kirchner (1880–1938). German painter, sculptor, designer and graphic artist; born at Aschaffenburg, Franconia. Active after 1905 in Munich, Berlin and Längmatt, Switzerland. There is a large collection of his work in Stuttgart.

Semyon Davidovitch Kirlian (c. 1906–78). Russian inventor. Based in Krasnodar, Soviet Caucasia.

Paul Klee (1879–1940). Swiss painter, musician and graphic artist; born at Münchenbuchsee. Active largely in Munich 1898–1921, Bauhaus master at Weimar and Dessau 1921–31, then based in Berne. Publications include *Pädagogisches Skizzenbuch*, Munich 1925. There are large collections of his work in Berne (Klee Foundation and Felix Klee Collection).

Yves Klein (1928–62). French painter, sculptor, musician and light artist; born at Nice, Alpes-Maritimes. Based in Tokyo 1952–53 and then largely in Paris.

Rockne Krebs (1938). American sculptor, graphic artist and light artist; born in Kansas City, Missouri. Based in Washington D.C. since 1964.

František Kupka (1871–1957). Czechoslovak painter, spiritist and graphic artist; born at Opočno, Bohemia. Active in Vienna 1891–95 and then largely in Paris. There are large collections of his work in Paris (National Museum of Modern Art) and Prague (Kupka Museum).

William Kurtz (1834–1904). American graphic artist and photographer; born in Germany. Based in New York City 1858–60 and after 1865.

Edwin Herbert Land (1909). American physicist and inventor; born in Bridgeport, Connecticut. Founded the Polaroid Corporation in Cambridge, Massachusetts, 1937.

Bernard Leach (1887–1979). British ceramicist; born in Hong Kong. Based in Tokyo 1909–20 and then at St. Ives, Cornwall. Publications include *A Potter's Book*, London 1940.

Jacob Christoph Le Blon (1670–1741). Saxon engraver; born at Frankfurt-am-Main. Active after 1696 in Rome, Amsterdam, London and Paris. Publications include *Coloritto, or, The Harmony of Colouring in Painting*, London 1735.

Fernand Léger (1881–1955). French painter, sculptor and film-maker; born at Argentan, Normandy. Active largely in Paris after 1903. There is a Fernand Léger Museum at Biot, Côte d'Azur.

Leonardo da Vinci (1452–1519). Tuscan painter, sculptor, anatomist, inventor and architect. Active largely in Milan 1482–1513. Collected theories published as *Trattato della pittura*, Milan 1651.

Roy Lichtenstein (1923). American painter, sculptor and graphic artist; born in New York City. Based in Cleveland, Ohio, 1949–57 and then South Hampton, New York.

Morris Louis (Morris Louis Bernstein) (1912–62). American painter; born in Baltimore, Maryland. Based in Baltimore 1933–47 and then Washington D.C.

Jean Lurçat (1892–1966). French painter, designer and graphic artist; born at Bruyères, Vosges. Active after 1912 largely in Paris and Switzerland; widely travelled. Publications include *Tapisserie française*, Paris 1950.

Max Lüscher (1923). Swiss psychologist; born in Basel. Based in Basel since 1946. Publications include *Klinischer Test zur Personlichkeitdiagnostik*, Basel 1948.

Len Lye (1901). American film-maker, sculptor and light artist; born in Christchurch, New Zealand. Working in Australia, London and (since 1946) New York City.

John McCracken (1934). American painter and sculptor; born in Berkeley, California. Working in Venice, California, and Las Vegas, Nevada, since 1965.

Norman McLaren (1914). Canadian film-maker; born in Stirling, Scotland. Based in London, Ottawa and (since 1956) Montreal.

Theodore Harold Maiman (1927). American physicist; born in Los Angeles. Working in Malibu and Los Angeles, California, since 1955.

Frank Josef Malina (1912). American aeronautical engineer, geophysicist, editor, graphic artist and light artist; born at Brenham, Texas. Working in Pasadena, California, 1936–46 and then Paris and Boulogne-sur-Seine.

Henri Matisse (1869–1954). French painter, sculptor, designer and graphic artist; born at Le Cateau-Cambrésis, Nord. Active after 1890 largely in Paris, Nice and Vence, Alpes-Maritimes; widely travelled. Publications include *Notes d'un peintre*, Paris 1908. There are large collections of his work in Nice (Matisse Museum), Baltimore (Museum of Art) and Merion, Pennsylvania (Barnes Foundation).

James Clerk Maxwell (1831–79). Scots physicist and mathematician; born in Edinburgh. Based in Aberdeen 1856–60 and then largely in Cambridge. Publications include *On the Theory of Compound Colours*, London 1860 and *On a Dynamical Theory of the Electro-Magnetic Field*, London 1864.

Piet Mondrian (1872–1944). Dutch painter; born in Amersfoort. Active in Amsterdam, Paris and New York City. Publications include *Le néo-plasticisme*, Paris 1920. There are large collections of his work in The Hague (Gemeentemuseum) and New York (Harry Holtzman Collection).

Claude Monet (1840–1926). French painter; born in Paris. Active largely in Paris 1862–83 and then at Giverny, Eure. There are large collections of his work in Boston and in Paris (Musée Marmottan and Galerie du Jeu de Paume).

Edvard Munch (1863–1944). Norwegian painter and graphic artist; born at Löten, Hedmark. Active after 1885 in Paris, Berlin and then at Kragerö and Ekely, Norway. There is a Munch Museum in Oslo.

Albert Henry Munsell (1858–1918). American painter and inventor; born in Boston, Massachusetts. Based in Boston after 1881. Publications include *A Color Notation*, New York and London 1905 and *Color Balance*, Boston 1913.

Barnett Newman (1905–70). American painter and sculptor; born in New York City. Active largely in New York after 1922.

Sir Isaac Newton (1642–1727). English natural philospher and mathematician; born at Woolsthorpe, Lincolnshire. Based in Cambridge 1661–95 and then largely in London. Publications include *New Theory about Light and Colours*, London 1672 and *Opticks*, London 1704.

Kenneth Noland (1924). American painter; born at Asheville, North Carolina. Working since 1949 in Washington D.C. and South Shaftesbury, Vermont.

Emil Nolde (Emil Hansen) (1867–1956). German painter and graphic artist; born at Nolde, Denmark. Active after 1887 in Dachau, Munich, Alsen, Berlin and Seebüll, Schleswig-Holstein; widely travelled. There is a Nolde Foundation at Seebüll.

Friedrich Wilhelm Ostwald (1853–1932). German physical chemist; born in Riga, Latvia. Based in Riga, Leipzig and Saxony. Nobel Prizewinner 1909. Publications on colour include *Die Farbenfibel*, Leipzig 1916 (republished 1917 with colour samples) and *Die Harmonie der Farben*, Leipzig 1918.

Nam June Paik (1932). Korean composer, light artist and video artist; born in Seoul. Based in Tokyo 1952–58 and then Cologne, New York City and Boston.

Sir William Henry Perkin (1838–1907). English industrial chemist; born in London. Founded the aniline dye industry at Greenford Green, Middlesex, 1857.

Zdeněk Pešánek (1896–1965). Czechoslovak sculptor, architect and light artist. Based in Prague. Publications include *Kinetismus*, Prague 1941.

Camille Pissarro (1831–1903). French painter and graphic artist; born at St. Thomas, West Indies. Active in Venezuela 1853 and then largely in Paris.

Antonio Pollaiuolo (c. 1432–98). Florentine painter, sculptor, goldsmith and engraver. Active in Florence after 1457.

Larry Poons (1937). American painter and musician; born in Tokyo. Working largely in New York City since 1958.

Johannes Purkinje (Jan Purkyně) (1787–1869). Bohemian physiologist and histologist; born at Lobkowitz. Based in Breslau 1823–50 and then Prague. Publications include *Beobachtungen und Versuche zur Physiologie der Sinne* (2 volumes), Berlin and Prague 1823–25.

Pythagoras of Samos (c. 572–c. 497 B.C.). Greek natural philosopher, astronomer and mathematician. Active in Crotona, Italy, after 532 B.C.

Robert Ridgway (1850–1929). American naturalist and ornithologist; born at Mount Carmel, Illinois. Based in Washington D.C. after 1880. Publications include *A Nomenclature of Colours for Naturalists*, Washington D.C. 1886 (revised 1912 as *Color Standards and Color Nomenclature*).

Bridget Riley (1931). English painter and graphic artist; born in London. Working largely in London since 1961.

Diego Rivera (1886–1957). Mexican painter and designer; born at Guanajuato. Active in Paris 1911–21 and then largely in Mexico City.

Ogden Nicholas Rood (1831–1902). American physicist; born at Danbury, Connecticut. Based in New York City after 1863. Publications include *Modern Chromatics*, New York and London 1879 (republished 1881 as *Students' Text-Book of Color*).

Georges Rouault (1871–1958). French painter, designer and graphic artist; born in Paris. Active largely in Paris after 1891. There is a large collection of his work in Paris (National Museum of Modern Art).

Robert D. Routh (1921). American photographer and writer; born in Chicago. Working in Anaheim and Long Beach, California. Publications include *Photographics*, Los Angeles 1976.

Philipp Otto Runge (1777–1810). German painter; born in Pommern. Active in Dresden 1801–4 and then Hamburg. Published *Die Farbenkugel*, Hamburg 1810.

John Ruskin (1819–1900). English writer and graphic artist; born in London. Widely travelled. Publications include *Modern Painters* (5 volumes), London 1843–60 and *The Elements of Drawing*, London 1857.

Peter Sedgley (1930). English painter, sculptor and light artist; born in London. Working in London 1963–71 and then West Berlin.

Georges Seurat (1859–91). French painter and graphic artist; born in Paris. Principal painter of Neo-Impressionism in Paris after 1881.

W. Christian Sidenius (1923). American designer and light artist; born in East Rutherford, New Jersey. Based at Sandy Hook, Connecticut, since 1963.

Rudolph Steiner (1861–1925). Austrian philosopher, educator, painter and architect; born at Kraljevee, Serbia. Based in Weimar 1890–97, then Berlin and Dornach, Switzerland. Publications include *Das Wesen der Farben*, Dornach 1921.

Frank Stella (1936). American painter and designer; born at Malden, Massachusetts. Working largely in New York City since 1958.

Louis-Jacques Thenard (1777–1857). French chemist; born at La Louptière, Seine. Based in Paris after 1794.

Louis Comfort Tiffany (1848–1933). American painter and designer; born in New York City. Founded the Tiffany Glass and Decorating Company in New York 1878.

Titian (Tiziano Vecelli) (c. 1487–1576). Venetian painter; born at Piève di Cadore. Active in Venice after 1508, then Bologna, Rome and Augsburg.

Joseph Mallord William Turner (1775–1851). English painter and graphic artist; born in London. Active in London 1789–1819 and then widely travelled. There are large collections of his work in London (British Museum and Tate Gallery).

Stan VanDerBeek (1927). American film-maker, video artist and graphic artist; born in New York City. Working largely in New York City and Stony Point, New York, since 1954.

Giorgio Vasari (1511–74). Tuscan painter, sculptor, architect and writer; born in Arezzo. Active in Arezzo, Rome and Florence. Founded the Academy of Design in Florence 1563. Published *Le vite de' più eccellenti architetti, pittori, e scultori italiani*, Florence 1550 (enlarged edition 1568).

Louis-Nicholas Vauquelin (1763–1829). French chemist; born at Saint-André-d'Hébertot, Normandy. Based in Paris after 1794.

Tom Wesselmann (1931). American painter and sculptor; born in Cincinnati, Ohio. Working largely in New York City since 1956.

Thomas Wilfred (Richard Lövström) (1889–1968). American musician and light artist; born in Copenhagen. Based in New York after 1916. Founded the Art Institute of Light in New York 1930.

Andrew Wyeth (1917). American painter; born at Chadds Ford, Pennsylvania. Based at Chadds Ford, Pennsylvania.

Thomas Young (1773–1829). English physician, physicist and Egyptologist; born at Milverton, Somerset. Based in Cambridge 1797–99 and then London. Publications include *A Course of Lectures on Natural Philosophy and the Mechanical Arts*, London 1807.

Vladimir Kosma Zworykin (1889). American physicist, electrical engineer and inventor; born in Mourom, Russia. Based in Leningrad, Paris, Pittsburg and Camden, New Jersey. Publications include *Television in Science and Industry*, New York and London 1958.

APPENDIX 3

List of Pigments

Pigment *permanence* (lightfastness or colurfastness) refers to the property of a colouring agent to resist a modification of its light-absorption characteristics. Materials change colour as a result of alterations in their molecular structure, such changes occurring variously as a consequence of ageing, improper storage, contact with water or damaging chemicals, or exposure to intense light or extremes of temperature.

Daylight and blue light are potentially more harmful than red light of equal duration and intensity; incandescent filament lamps or 'warm white' fluorescent lamps are therefore safer for lighting than the bluer 'cool white' sources. Commercial pigments which have resisted fading after 600 hours of exposure to direct sunlight justify the term 'absolutely permanent'; such pigments include most raw earths, natural mineral colours and all carbon-based black pigments.

Intermixing pigments often has an adverse effect on pigment permanence. This is especially so in mixtures of chemically incompatible colours, such as a mixture of a chromium pigment with one containing sulphur. Tinting a pigment with extender (commonly blanc fixe or calcium carbonate) invariably weakens its permanence, most especially in very pale tints. Cobalt blue, Chrome yellow and Flake white are particularly susceptible to damage by water or a damp atmosphere. Prussian blue is bleached by alkaline attack whereas both Genuine and French ultramarine are bleached by acids. Such damage is almost always irreversible though Lithopone (zinc sulphide and blanc fixe) exhibits the *photochromic* property of darkening as the result of prolonged exposure to light and regaining its whiteness during periods of darkness. Prussian blue also exhibits photochromism; and Chrome yellow reddens if heated but returns to its original colour on cooling.

Pigment *opacity* refers to 'hiding power' or the capacity of pigment particles in suspension to scatter incident light in order to obscure the ground to which a layer of paint or printing ink has been applied. The general rule here is that, when light is passing from a transparent material of low refractive index into one of higher refractive index, the amount of light reflected at the boundary will be greater the greater is the difference between the two indices. The property of relative opacity is therefore largely dependent on the difference between the refractive index of the pigment and the refractive index of its binder, rather than the concentration of pigment particles contained in the paint or ink layer. (The refractive index of a material indicates its relative optical 'density'; it is calculated by dividing the velocity of light in a given medium by its standard velocity in a vacuum. The ratio is normally quoted as an average for all spectral colours though the index for the shorter wavelengths is slightly higher than that for the longer wavelengths.)

The refractive indices of most oil and polymer vehicles fall within the range 1·4 to 1·6. Titanium dioxide (Titanium white) is the most opaque white pigment by virtue of a refractive index (2·72) which differs considerably from that of such a binder. At the other extreme, alumina hydrate, which has a refractive index (1·54) little different from that of an oil binder, appears highly transparent if suspended in oil.

A pigment suspended in a water-soluble vehicle will appear to lighten as the vehicle dries: the spaces occupied by water (refractive index 1·33) are being replaced gradually by spaces occupied by air (refractive index 1·00) and a greater proportion of the incident light is being reflected. Conversely, the refractive index of linseed oil (1·48) increases as it dries and pigments suspended in an oil vehicle will appear to darken. The more transparent the pigment the more it will be affected by this change; hence a pigment such as Raw sienna, which is itself highly permanent, will in time appear to darken if suspended in oil.

There follows an alphabetical list of the most common pigments currently in use by artists.

Alizarin crimson. An artificial lake derived from coal tar (dihydroxy anthraquinone). *Poisonous.* Moderately permanent. Transparent. Not to be mixed with pigments containing lead.

Bone black (Ivory black). Carbon and calcium phosphate, formerly obtained by charring ivory chips in a closed retort; now bone (animal charcoal) is used. Nonpoisonous. A highly permanent 'cool' black which mixes well. Fairly transparent.

Cadmium red, scarlet, orange, yellow and lemon. Cadmium sulphide coprecipitated with blanc fixe (artificial barium sulphate), with some selenide added for reds and oranges. *Poisonous.* Permanent. Opaque. An indispensible range of pigments the inferior grades of which discolour if mixed with pigments containing copper or lead.

Carbon black (Gas black, Furnace black or Channel black). Pure natural carbon, obtained by the incomplete combustion of methane (marsh gas). Nonpoisonous. Highly permanent. Opaque.

Chrome orange, yellow and lemon. Lead chromates. *Poisonous.* Not permanent. Opaque. Chrome pigments tend to fade in direct sunlight; they discolour if mixed with pigments containing sulphur and are advisedly replaced by the Cadmium and Hansa yellows.

Chromium oxide green. Anhydrous chromium oxide. *Poisonous.* Highly permanent. Highly opaque.

Cobalt blue. Combined oxides of cobalt and aluminium. Nonpoisonous. Permanent. Transparent.

Flake white (White lead). Lead carbonate, often with zinc oxide added. *Poisonous.* Highly opaque. A permanent 'warm' white which tends to yellow with age and form black lead sulphide if mixed with pigments containing sulphur or exposed to a sulphurous industrial atmosphere. A pigment advisedly replaced by Titanium white or Zinc white.

French ultramarine. A calcined compound of silica, alumina, soda ash and sulphur. Nonpoisonous. A permanent pigment of exceptional spectral purity. Fairly transparent. Discolours if mixed with pigments containing lead or copper.

Green earth (Terre verte or Veronese green). Natural green earth (glauconite) rich in silicates of iron, manganese, aluminium and potassium. Nonpoisonous. Highly permanent. Transparent. Not to be mixed with pigments containing lead.

Hansa orange G, yellow 5G and lemon 10G. Artificial lakes derived from coal tar (diazotised toluidine or nitraniline). *Poisonous.* Highly permanent. Transparent.

Ivory black. See Bone black.

Lampblack. Pure natural carbon: soot collected by the incomplete combustion of oil or coal-tar products. Nonpoisonous. A highly permanent 'cool' black. Highly opaque. India ink consists of Lampblack dispersed with gum binder in water.

Light red (English red or Burnt ochre). Natural yellow earth, rich in hydrous oxides of iron and silicon, which has been calcined (reduced to powder by heat). Nonpoisonous. Highly permanent. Opaque.

Mars red, orange, yellow, violet, brown and black. Artificial iron oxides of the same chemical constitution as natural earths. Nonpoisonous. Highly permanent. Opaque.

Phthalocyanine blue and green. Artificial lakes derived from coal tar (copper phthalocyanine) with chlorine added for greens. *Poisonous.* Highly permanent. Opaque. Proprietary names for the pigments include Bocour blue and green, Monastral blue and green, Monestial blue and green, Thalo blue and green and Windsor blue and green.

Quinacra red, magenta and violet. Artificial lakes derived from coal tar (linear quinacridone). *Poisonous.* Highly permanent. Transparent.

Raw sienna. Natural yellow-brown earth rich in hydrous oxides of iron and manganese. Nonpoisonous. Highly permanent. Fairly transparent. Calcined or **Burnt sienna** is a permanent dark-brown pigment reduced to powder by heat.

Raw umber. Natural brown earth rich in hydrous oxides of iron and manganese. Nonpoisonous. Highly permanent. Fairly transparent. Calcined or **Burnt umber** is a permanent reddish-brown pigment reduced to powder by heat.

Red ochre (Indian red or Venetian red). Natural red earth rich in hydrous iron oxide (hematite). Nonpoisonous. Highly permanent. Highly opaque.

Titanium white. Titanium dioxide, with some zinc oxide and blanc fixe added. Nonpoisonous. A permanent, neutral white. Highly opaque.

Ultramarine blue (artificial). See French ultramarine.

Vandyke brown (Cologne earth). Natural brown coal (peat or lignite). Nonpoisonous. Not permanent. Transparent.

Vermilion. Mercuric sulphide. *Poisonous.* Moderately permanent. Highly opaque. Tends to darken if over-exposed to direct sunlight or if mixed with pigments containing lead or sulphur. A pigment advisedly replaced by Cadmium red.

Viridian (Verte émeraude). Hydrous chromium oxide, or a compound of copper phthalocyanine and chlorine. *Poisonous.* Highly permanent. Transparent.

Yellow ochre. Natural yellow earth (limonite) rich in hydrous oxides of iron, aluminium and silicon. Highly permanent. Highly opaque.

Zinc white (Chinese white). Zinc oxide. A highly permanent 'cool' white which, unlike Flake white, does not discolour. Nonpoisonous. Fairly transparent.

Zinc yellow. Zinc chromate. *Poisonous.* Moderately permanent. Fairly transparent.

APPENDIX 4

Select Bibliography

PART I

Faber Birren (1969), *Light, Color and Environment.* New York: Van Nostrand Reinhold Company.

Ralph M. Evans (1948), *An Introduction to Color.* New York: John Wiley and Sons; London: Chapman and Hall.

Johann Wolfgang von Goethe and Rupprecht Matthaei, editor (1961), *Goethe's Color Theory.* New York: Van Nostrand Reinhold Company; London: Studio Vista. New edition 1971.

Ralph Norman Haber and Maurice Herschenson (1973), *The Psychology of Visual Perception.* New York: Holt, Reinhardt and Winston. London edition 1974.

Hal Hellman (1967), *The Art and Science of Color.* New York: McGraw-Hill Book Company.

Dean B. Judd and Günter W. Wyszecki (1952), *Color in Business, Science, and Industry.* New York and London: John Wiley and Sons. Second edition 1967.

Wassily Kandinsky (1912) and Robert Motherwell, editor (1947), *Concerning the Spritual in Art.* New York: George Wittenborn.

David Katz (1935), *The World of Colour.* London: Kegan, Paul, Trench, Trubner and Company.

Conrad D. Mueller and Mae Rudolph, editors (1966), *Light and Vision.* New York and London: Time-Life Books. Pocket edition 1969.

John N. Ott (1973), *Health and Light.* Old Greenwich, Connecticut: Devin-Adair Company; New York: Pocket Books, 1976.

Maurice H. Pirenne (1970), *Optics, Painting and Photography.* London: Cambridge University Press.

Richard C. Teevan and Robert C. Birney, editors (1961), *Color Vision.* New York: D. Van Nostrand Reinhold Company.

Patrick Trevor-Roper (1970), *The World of Blunted Sight.* London: Thames and Hudson.

William D. Wright (1967), *The Rays are not Coloured.* London: Adam Hilger.

Günter W. Wyszecki and Walter S. Stiles (1967), *Color Science.* New York and London: John Wiley and Sons.

PART II

Ralph M. Evans (1959), *Eye, Film and Camera in Color Photography*. New York and London: John Wiley and Sons.

Fred Gettings (1971), *Polymer Painting Manual*. London: Studio Vista.

Colin Hayes (1978), *The Complete Guide to Painting and Drawing Materials and Techniques*. Oxford: Phaidon Press.

Robert W. G. Hunt (1957), *The Reproduction of Colour in Photography, Printing and Television*. King's Langley, Hertfordshire: Fountain Press. Third edition 1975.

Adrian Bernard Klein (1937), *Coloured Light: An Art Medium*. London: Technical Press.

Frank J. Malina, editor (1974), *Kinetic Art: Theory and Practice*. New York: Dover Publications.

Thelma R. Newman (1964), *Plastics as an Art Form*. Philadelphia: Chilton Book Company; London: Sir Isaac Pitman and Sons. Revised edition 1972.

Christopher. R. G. Reed (1969), *Principles of Colour Television Systems*. London: Sir Isaac Pitman and Sons.

Joseph F. Robinson (1975), *Videotape Recording, Theory and Practice*. London: Focal Press.

Joyce Storey (1978), *Dyes and Fabrics*. London: Thames and Hudson.

Kit van Tulleken, editor (1976), *The Techniques of Photography*. Amsterdam: Time-Life International.

Michael Wenyon (1978), *Understanding Holography*. Newton Abbot, Devonshire: David and Charles.

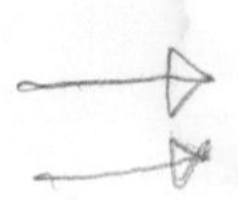

Laurie C. Young (1973), *Materials in Printing Processes*. London: Focal Press.

Gene Youngblood (1970), *Expanded Cinema*. London: Studio Vista.

PART III

Josef Albers (1963), *Interaction of Color*. New Haven, Connecticut: Yale University Press. Revised paperbound edition 1975.

Cyril Barrett (1970), *Op Art*. London: Studio Vista.

Michel-Eugène Chevreul (1839) and Faber Birren, editor (1967), *The Principles of Harmony and Contrast of Colors*. New York: Reinhold Publishing Corporation.

Anton Ehrenzweig (1967), *The Hidden Order*. London: Weidenfeld and Nicholson; Los Angeles and Berkeley: University of California Press. Second edition 1971.

Frans Gerritsen (1973), *Theory and Practice of Colour*. London: Studio Vista.

Hermann von Helmholtz (1867) and James P. C. Southall, editor (1962), *Treatise on Physiological Optics*. New York: Dover Publications; London: Constable and Company.

William Innes Homer (1964), *Seurat and the Science of Painting*, Cambridge, Massachusetts: M.I.T. Press.

Johannes Itten (1961), *The Art of Color*. New York: Reinhold Publishing Corporation. New edition 1973.

Andreas Kornerup and J. H. Wanscher (1961), *Methuen Handbook of Colour*. London: Eyre Methuen. Third edition, revised, 1978.

Albert H. Munsell (1905) and Faber Birren, editor (1969), *A Grammar of Color*. New York: Van Nostrand Reinhold Company.

Wilhelm Ostwald (1916) and Faber Birren, editor (1969), *The Color Primer*. New York: Van Nostrand Reinhold Company.

Paul Renner (1947), *Color: Order and Harmony*. New York: Reinhold Publishing Corporation; London: Studio Vista. English edition 1964.

Ogden N. Rood (1879) and Faber Birren, editor (1973), *Modern Chromatics*. New York: Van Nostrand Reinhold Company.

John Ruskin (1857), *The Elements of Drawing*. New York: Dover Publications; London: Constable and Company. Reprint 1971.

William D. Wright (1944), *The Measurement of Colour*. London: Adam Hilger. Fourth edition 1969.

Index

absorption of light 14–17, 73, 74, 96, 118, 149
Academy of Sciences, Paris 107
acrylic resin 68, 69
acupuncture 32
adaptation 26–28, 85
additive colour mixing 10–12, 15, 16, 48, 50, 90, 116, 133
additive primary colours 10, 11, 14, 15, 24, 43, 49, 50, 56, 90, 116, 136
aerial perspective 121, 122, 123, 124
aerosol paint 68
after-image 109, 110, 111, 114
Agis, Maurice 69, 70, 139
Albers, Josef 71, 113, 116, 124, 125, 139
Alhambra Palace 111
Alizarin 61
Alizarin crimson 99, 150
alkyd resin 68, 69
alumina hydrate 61, 150
aluminium 45, 59, 60, 71, 89, 127, 151, 152
amplitude 19, 21, 54, 133
aniline 61
anthraquinone 150
anticerne 113
antimony 49
Antonioni, Michelangelo 80, 139
aquarelle 67
aqueous paint 66, 67
arclight 41–43, 53
argon 39
argon lamp 42
argon laser 46, 77
arsenic 59
Aubertin, Bernard 37, 139
aura 30, 31
Azurite 59

Baird, John Logie 48, 53, 139
Bakelite 68, 69
batik 61, 66
Bauhaus 71, 124, 139, 143
beeswax 65
beetroot 59
Belson, Jordan 40, 80, 81, 139
Benyon, Margaret 54, 57, 139
bismuth 42
black contour 112, 113
Blanc, Charles 118, 139
blanc fixe 61, 149, 150, 152
Bocour blue and green 152
body colour 67
Bone black 150
Bonnard, Pierre 123, 124, 139
Botticelli, Sandro 65, 139
Bouguer, Pierre 26, 85, 139
Bourges Cathedral 71
brazilwood 59
Breer, Robert 80, 81, 110, 111, 139
brightness *see* luminosity
Bristol Blue glassware 59
buckthorn 60
Burnt ochre 151
Burnt sienna 99, 152
Burnt umber 152
Burrell, Fred 79, 140
butane 38

cadmium 59
Cadmium reds 150, 151
Cadmium yellow 99, 150, 151
cadmium vapour lamp 42
calcium 38, 43, 70, 150
camera, film 4, 80, 81
  photographic 54, 78, 80
  television 49–53, 77
camomile 60
candela 86, 88, 134
candlelight 4, 37, 39, 85, 86
carbon 37, 58, 59, 149, 150, 151
carbon arc lamp 41
Carbon black 15, 151
carbon dioxide lamp 42
Carmine 59
casein 68
Castel, Louis-Bertrand 37, 99, 140
cathode-ray tube 49, 50
Cayce, Edgar 31, 140

cellulose 68, 77, 78, 81, 134
ceramic glaze 71, 72
cerium 59
Cézanne, Paul 67, 99, 113, 118, 129, 140
Chagall, Marc 70, 71, 140
Chalk white 98
champlevé enamel 71
Channel black 151
Chartres Cathedral 71
Chevreul, Michel-Eugène 6, 14, 17, 96, 100, 107, 109, 111, 112, 114, 116, 117, 119, 121, 140
chiaroscuro 121, 122, 123
Chinese white 152
chlorine 61, 152
chroma 104, 105
chromatic circle *see* colour wheel
chromaticity 89, 92, 94, 133
chromaticity diagram 90, 92, 93, 103
Chrome yellow 60, 99, 149, 151
chromium 45, 52, 59, 60, 149, 152
Chromium oxide green 151
Chryssa 42, 140
cinepainting 81
Cinnabar 59, 60
cloisonné enamel 71, 112
Cloisonism 112
closed-circuit television 51
coal tar 61, 68, 150, 151, 152
cobalt 52, 59, 151
Cobalt blue 60, 99, 149, 151
cochineal 59
coherent light 45, 53
cold-setting plastics 69
Cologne earth 152
colorimeter 90
colour atlas 96
colour blindness 25, 26
colour constancy 4–6, 16, 25
colourfastness 149
colourfulness *see* saturation
colour healing 32
colour matching 95, 98
colour pencil 66
colour separation 73–75, 76, 77, 78, 99
colour sphere 102
colour test 29
colour therapy 32, 33
colour vision deficiency 25, 26
colour wheel 100–102, 103, 118, 119
combustion 37
Commission Internationale de l'Éclairage 90, 91
complementary colour 15, 16, 49, 75, 79, 93, 100, 101, 107, 108, 109, 112, 116, 117, 133
cone vision 23–28
Conrad, Tony 114, 140
continuous-tone printing 76
copolymer paint 68
copper 38, 42, 59, 60, 61, 71, 127, 151
Cornwell-Clyne, Adrian 41, 140
coumarone-indene resin 68
coupler 78, 79
cranberry 59
Crimson 59
critical fusion frequency 114, 115
Cross, Lloyd 54, 57, 140
Cruz-Diez, Carlos 120, 140
Cubism 22, 109
cuttlefish 59

Dali, Salvador 54, 140
dark-adaptation 27, 28, 97
Davis, Ronald 69, 140
Degas, Edgar 66, 141
Delacroix, Eugène 99, 113, 117, 118, 141
Delaunay, Robert 6, 108, 109, 112, 128, 129, 141
Democritus 99, 141
dextrin 66, 67
dichroic filter 41, 74
Diesbach's paint manufactory 60
digital clock 43
discharge lamp 41, 42, 45, 114
Disney, Walt 80, 141
dispersion of light 8, 9, 41, 47, 56, 136
distemper 67, 68
Divisionism 118, 119
dominant wavelength 89, 92, 96, 97, 137
Dryer, Ivan 46, 47, 141
drying oil 63, 64, 68
Ducos du Hauron, Louis 73, 141
Dunning, George 80, 141
dye laser 46
dyer's plants 59, 60

earth pigments 58, 59, 149
Edison, Thomas Alva 39, 141
egg-tempera paint 63, 65, 67
egg-white 67
egg-yolk 63, 65
Egyptian blue 60

Eidophor 53
electroluminescence 42, 43, 53
electromagnetic energy *see* radiation
electronic engraving 76, 77
emulsion paint 68, 69
enamelling 70, 71, 112
encaustic paint 65, 66
English red 151
epoxy resin 69
exposure meter 88
eyeball 21, 22

Fauvism 113
Fechner, Gustav Theodor 103, 115, 116, 141
Feeley, Paul 113, 129, 141
Fénéon, Félix 118, 119, 141
Fernand Léger Museum 72
Field, George 99, 109, 141
filament lamp 38–40, 86, 149
film camera 4, 80
film projector 39, 41, 42
film speed 80, 114
Finsen, Niels 32, 141
firelight 4, 37, 38
Flake white 60, 99, 117, 149, 151, 152
flame colours 38
Flavin, Dan 44, 142
fluorescence 43–45, 134
fluorescent lamp 43, 44, 149
formaldehyde 68
Formica 69
four-colour printing 75
fovea 21, 23
French ultramarine 60, 99, 149, 151
frequency 19, 20, 51, 134
Furnace black 151

Gabor, Dennis 53, 142
Gas black 151
gas laser 46, 47
gaslight 37, 38, 85, 86
Gaudí, Antonio 71, 142
Gauguin, Paul 29, 71, 112, 142
gelatin 66, 67, 78, 134
gesso 65
glair 67
glass 8, 17, 59, 70, 71, 134
glass enamel 70–72
glass fibre 69
glass mosaic 70, 117
glauconite 151
glue size 65, 67
glycerol 68
glyptal resin 68
Gobelins, Royal Manufactory of the 61, 62, 96, 117
Goes, Hugo van der 64, 142
Goethe, Johann Wolfgang von 30, 33, 99, 109, 116, 142
Gogh, Vincent van 29, 142
gold 33, 59, 71
gouache 67
gravure printing 64, 76
Green earth 59, 99, 151
Guimet, Jean-Baptiste 60, 142
gum arabic 63, 66, 67, 71, 151
Guy's Hospital 32

Hadassah Medical Centre 71
haemin 61
half-tone printing 76–78, 119
halogen 38, 39, 74
Hamilton, Richard 77, 142
Hansa yellows 151
Harris, Moses 96, 99, 100, 109, 142
helium-and-cadmium laser 46
helium-and-neon laser 46
Helmholtz, Hermann von 24, 25, 142
hematite 152
hempseed oil 64
Hering, Ewald 100, 103, 142
hiding power 149, 150
host crystal 42, 45
hue 96, 97, 98, 100, 102, 103, 104, 105, 119, 134
Hughes Research Laboratories 45

illuminance *see* illumination
illumination 4, 25, 85, 88, 89, 134
Imperial Chemical Industries 61
Impressionism 6, 99, 108, 112, 118, 121
incandescence 37, 38, 134
incandescent lamp 38, 39
India ink 151
Indian red 152
Indigo 60, 61
infrared energy 21, 46, 97, 136
integral tripack film 78, 80
intensity of light 11, 38, 49, 85, 86, 88, 133, 134
interference pattern 54, 55
International Commission on Illumination 90, 91

International System of units 20, 87, 88
inverse square law 86
Iredell, Charles Edward 32, 142
iris flower 60
iron oxide 52, 59, 60, 151, 152
irradiation of colour 107, 111, 113, 117
isolation of colour 111, 113, 122
Itten, Johannes 71, 124, 143
Ivory black 150

Jenkins, Paul 67, 69, 143
Johns, Jasper 42, 66, 125, 143
Jones, Peter 69, 70, 143

Kandinsky, Wassily 29, 143
Kelly, Ellsworth 127, 143
Kepler, Johann 86, 143
kermes 59
kinescope 49
Kirchner, Ernst Ludwig 29, 143
Kirlian, Semyon 30, 143
Klee, Paul 66, 67, 71, 143
Klein, Adrian Bernard 41, 140
Klein, Yves 37, 143
Kodachrome 80
Krebs, Rockne 46, 47, 144
krypton laser 46, 47
Kupka, František 108, 116, 144
Kurtz, William 39, 73, 144

lake pigment 61
Lampblack 151
Land, Edwin H. 3, 4, 25, 120, 144
lapis lazuli 59
laser 32, 45–47, 53, 54, 56, 77
lavender 64
Leach, Bernard 71, 144
lead 49, 60, 113, 150, 151, 152
Le Blon, Jacob Christoph 8, 99, 144
lecithin 65
Léger, Fernand 6, 32, 70, 72, 112, 144
Le Mans Cathedral 71
Leonardo da Vinci 107, 122, 123, 124, 144
letterpress printing 64, 76
Lichtenstein, Roy 71, 77, 126, 144
light-adaptation 27, 28, 97
lightfastness 149
light meter 88
lightness 96, 97, 98, 102, 104, 117, 123, 135
Light red 151
lignite 152
limonite 152
linear perspective 121, 122, 124
linoxyn 63, 64
linseed oil 63, 64, 150
lithium 38
lithography 64, 76, 127
Lithopone 149
local colour 5, 67, 112, 123
logwood 60
Louis, Morris 69, 113, 144
lumen 87, 88, 89, 135
luminance 88, 89, 94, 96, 97, 135
luminance factor 89, 94
luminescence 42–45, 46, 115, 135
luminosity 96, 97, 135
luminous flux 87, 88, 135
luminous paint 44
Lurçat, Jean 62, 144
Lüscher, Max 29, 30, 144
lux 88, 134
Lyceum Theater 39
Lye, Len 80, 81, 144

McCracken, John 69, 145
McLaren, Norman 110, 111, 145
Madder 59, 99
magnesium 15, 59, 89
Maiman, Theodore H. 45, 145
Malachite 59
Malina, Frank J. 40, 145
manganese 42, 59, 151, 152
Mars pigments 152
matching stimuli 90
Matisse, Henri 29, 113, 119, 145
Mauve 61
Maxwell, James Clerk 3, 73, 90, 92, 116, 145
Maxwell's disc 116
Maxwell's triangle 90, 92
mercury 59, 152
mercury vapour lamp 42, 43
methane 151
methyl methacrylate 68
Metz Cathedral 71
milkwort 60
mineral pigments 58, 59, 149
mixture of lights *see* additive mixing
mixture of pigments *see* subtractive mixing
moiré pattern 76
Monastral blue and green 61, 152
Mondrian, Piet 127, 145
Monestial blue and green 152
Monet, Claude 6, 121, 126, 145
monochromatic light 45, 54, 90, 91, 98

monomer 67, 68
moonlight 28
mordant 60
Morrison Planetarium 40
mosaic glass 70, 117
motion-picture camera 4, 80
Multiplex holography 54, 57
Munch, Edvard 29, 66, 110, 145
Munsell, Albert H. 100, 103, 104, 116, 145
Munsell's colour system 104–106
murex shellfish 58
Museum Haus Lange 37
Museum of Modern Art, New York 40

Nabis 112
negative-colour film 79
Neighborhood Playhouse 40
Neo-Impressionism 119, 147
neon 42, 46
Newman, Barnett 30, 69, 145
Newton, Isaac 8, 9, 10, 19, 100, 145
Nice University 70
nickel 52, 59
nit 88, 89, 135
nitraniline 151
nitrogen 39
Noland, Kenneth 69, 113, 125, 146
Nolde, Emil 29, 67, 123, 146
Nôtre-Dame de Toute Grâce, Plateau d'Assy 70, 113

oil burner 85, 86
oil crayon 66
oil paint 63, 64, 119, 150
oil pastel 66
Omni International Building 46
optical mixing, by disc 115, 116, 117
  by mosaic 50, 76, 115, 117–119
optic nerve 22, 23
orchil 59
Orpiment 59
Ostwald's colour system 103, 104
Ostwald, Wilhelm 100, 103, 116, 146
oxidation 64, 78, 79
oxygen 33, 37, 39, 41, 64

Paik, Nam June 48, 146
Parliament Building, Jerusalem 70
pastel 66, 109, 118
Peach black 99
peat 152
Perkin, William Henry 61, 146
permanence 149, 150
persistence of vision 114
Perspex 69, 120
Pešánek, Zdeněk 40, 146
phenolic resin 68, 69
phosphorescence 43, 44, 50, 135
phosphor triads 50, 51
photochromism 149
photoengraving 73
photographic development 55, 74, 78, 79
photographic emulsion 78, 79
photosensitive chemicals 74, 78
photometer 85, 86
photopic vision 27
Phthalocyanine pigments 61, 152
picture tube 49
pineal gland 33
Pissarro, Camille 118, 119, 146
pituitary gland 33
plasticity 67, 69
platinum filament lamp 39
platinum furnace 86, 88
Plexiglas 69
Plumbicon 49
Pointillism 118
Pollaiuolo, Antonio 64, 146
polyester resin 69
polyether resin 69
polyethylene 69
polymerisation 67, 68
polymethyl methacrylate 68, 69
polypropylene 69
polystyrene 69
Polythene 69
polyvinyl acetate 68, 69
polyvinyl chloride 68, 69, 70
Poons, Larry 110, 146
poppyseed oil 64
portapak 53
poster colour 67, 80
posterisation 80
potassium 38, 59, 60, 61, 151
pot-metal glass 71
Pre-Raphaelite Brotherhood 117
primary colours of light *see* additive primary colours
primary colours of pigment *see* subtractive primary colours
projection of light 3, 10, 39–41, 47, 53, 79
propane 38
Prussian blue 60, 99, 149

psychological primary colours 100, 136
psychophysics 97, 103
purity 16, 17, 89, 90, 92, 96, 97, 136
Purkinje, Johannes 27, 28, 146
Pythagoras 32, 146

Quinacra reds 152
quinacridone 61, 152

radiation 136
Radio Corporation of America 48
radium 44
ragweed 60
rainbow 8
rainbow hologram 56, 57
Raw sienna 59, 99, 150, 152
Raw umber 59, 152
Realgar 59
Red ochre 59, 98, 99, 152
reflectance 89
reflection of light 14–18, 65, 89, 136, 149, 150
reflection print 56, 79
refraction of light 8, 150
refractive index 149, 150
retina 21–26, 33, 49
reversal-colour film 78, 79
Ridgway, Robert 96, 146
Riley, Bridget 120, 147
Rivera, Diego 66, 71, 147
rod vision 23, 24, 27, 28
Rood, Ogden N. 100, 103, 111, 116, 118, 119, 121, 123, 147
Rouault, Georges 71, 112, 147
Rouen Cathedral 126
Routh, Robert D. 79, 80, 147
Royal Institution 3, 115
Royal Society 9
rubidium 46
ruby laser 45
Runge, Philipp Otto 99, 100, 102, 103, 147
Ruskin, John 6, 31, 107, 117, 119, 147

saccade 22, 111
Sacré-Coeur, Audincourt 112
S. Marco, Venice 71
Sta. Sophia, Instanbul 70
S. Sulpice, Paris 118
safelight 16, 18
safflower 59
saffron 60
saturation 96, 97, 98, 99, 102, 104, 118, 136
scattering of light 18, 66, 122, 137, 149
scenic colour 67
scotopic vision 27
secondary colours, of light 12
of pigment 12, 101
Sedgley, Peter 40, 45, 125, 147
selenium 49, 59
separation filter 73, 74, 75
Sepia 59
serigraphy 68, 76, 77
setting plastics 69
Seurat, Georges 118, 119, 147
shade 97, 102
shadow mask 50
shellac 66, 68
Sidenius, Christian 40, 147
silica 38, 70, 151
silicon 59, 151, 152
silver 33, 42, 45, 71, 78, 89
silver halide 78
simultanéisme 109
simultaneous contrast 107, 108, 109
size 66, 67
slip 71, 72
smaltini 59
sodium 38, 42, 59, 70
sodium vapour lamp 42
solarisation 80
solid angle 87, 88
spectral sensitivity 26, 27, 97
spectrum 8, 21, 37, 96, 98, 100, 136
spectrum locus 91, 92
speed of light *see* velocity of light
spike oil 64
spray paint 68
stained glass 43, 59, 70, 111, 112
starch gum 66
steel 71
Steiner, Rudolph 32, 147
Stella, Frank 45, 69, 113, 127, 128, 147
strontium 38
subtractive colour mixing 12, 13, 17, 66, 73–75, 76, 98, 137
subtractive primary colours 12, 73, 75, 99, 101, 127, 136
successive contrast 109, 110, 111
sulphur 59, 149, 151, 152
sun-tanning 21

tapestry 58, 62, 96, 107, 117
television camera 49, 51, 52, 53, 77
television projector 41, 53

television receiver 43, 49–51, 52, 53
tellurium 38, 59
tempera paint 63, 65, 67
Terre verte 151
tertiary colour 101
tetrachome palette 98
Thalo blue and green 152
Thenard, Louis-Jacques 60, 147
thermoplastics 69, 70
thermosetting plastics 69
thorium 44
three-colour pigment vision 24
three-colour printing 73–75
Tiffany, Louis Comfort 39, 112, 147
tin 59, 60
tint 97, 102, 119, 149
titanium 59, 150
Titanium white 150, 151, 152
Titian 59, 99, 148
toluidine 151
tone 75, 76, 96, 135
toning 79
transmission of light 14–18, 74, 78, 137
transmittance 89
transparency print 3, 74, 77, 78, 79, 137
trichromatic printing *see* three-colour printing
tripack film 78, 79, 80
tristimulus values 94
tungsten filament lamp 38–41, 149
tungsten halogen lamp 39
Turner, J. M. W. 5, 67, 148
Tyrian purple 58

Ultramarine blue 59, 60, 99, 149, 152
ultraviolet energy 21, 32, 43, 44, 97, 136
urea-formaldehyde 69

value 104, 105
vanadium 59
VanDerBeek, Stan 48, 148
Vandyke brown 152
Vasari, Giorgio 107, 122, 148
Vauquelin, Louis-Nicholas 60, 148
velocity of light 8, 19, 54, 137, 150
Venetian red 152
Verdigris 60
Vermilion 60, 99, 152
Veronese green 151
Verte émeraude 152
videocassette 52
videodisc 52
videotape recording 48, 52, 53
Vidicon 49
Vine black 98
vinyl resin 63, 68, 69, 70
Viridian 99, 152
visual mixing *see* optical mixing
visual purple 24
vitamin A 24, 33
vitamin D 21
vitreous enamel 70–72

walnut oil 64
watercolour paint 66, 67, 112
water-soluble paints 66, 67, 68, 150
wavefront reconstruction 53, 55, 56
wavelength 19, 20, 21, 25, 26, 37, 89, 90, 98, 137
wave theory of light 19, 20, 54
wax crayon 65, 66
weld 60
Wesselmann, Tom 42, 48, 69, 70, 148
WGBH-TV 48
white contour 113
White lead 151
Wilfred, Thomas 39, 40, 148
Windsor blue and green 152
woad 60
Wyeth, Andrew 65, 67, 148

xenon lamp 42, 45, 77
X-ray energy 136

Yale University 113
Yellow ochre 59, 98, 99, 152
Young, Thomas 4, 19, 24, 25, 109, 115, 116, 148

zinc 42, 44, 149, 151, 152
zinc vapour lamp 42
Zinc white 151, 152
Zinc yellow 152
Zworykin, Vladimir 48, 148